OXFORD CLASSIC
THE PHYSICAL S

QUANTUM THEORY OF SOLIDS

BY
R. E. PEIERLS

CLARENDON PRESS • OXFORD

OXFORD
UNIVERSITY PRESS

Great Clarendon Street, Oxford OX2 6DP
Oxford University Press is a department of the University of Oxford.
It furthers the University's objective of excellence in research, scholarship,
and education by publishing worldwide in
Oxford New York
Athens Auckland Bangkok Bogotá Buenos Aires Calcutta
Cape Town Chennai Dar es Salaam Delhi Florence Hong Kong Istanbul
Karachi Kuala Lumpur Madrid Melbourne Mexico City Mumbai
Nairobi Paris São Paulo Shanghai Singapore Taipei Tokyo Toronto Warsaw
with associated companies in Berlin Ibadan

Oxford is a registered trade mark of Oxford University Press
in the UK and in certain other countries

Published in the United States
by Oxford University Press Inc., New York

© Oxford University Press, 1955

The moral rights of the author have been asserted
Database right Oxford University Press (maker)

First published 1955
Reprinted 1956, 1964, 1965, 1974
Published in the Oxford Classics Series 2001

All rights reserved. No part of this publication may be reproduced,
stored in a retrieval system, or transmitted, in any form or by any means,
without the prior permission in writing of Oxford University Press,
or as expressly permitted by law, or under terms agreed with the appropriate
reprographics rights organization. Enquiries concerning reproduction
outside the scope of the above should be sent to the Rights Department,
Oxford University Press, at the address above

You must not circulate this book in any other binding or cover
and you must impose this same condition on any acquirer

A catalogue record for this book is available from the British Library

Library of Congress Cataloging in Publication Data
Data available
ISBN 0 19 850781 X

Printed in Great Britain
on acid-free paper by
Redwood Books, Trowbridge, Wiltshire

PREFACE

THE quantum theory of solids has sometimes the reputation of being rather less respectable than other branches of modern theoretical physics. The reason for this view is that the dynamics of many-body systems, with which it is concerned, cannot be handled without the use of simplifications, or without approximations which often omit essential features of the problem. Nevertheless, the theory contains a good deal of work of intrinsic interest in which one can either develop a solution convincingly from first principles, or at least give a clear picture of the features which have been omitted and therefore discuss qualitatively the changes which they are likely to cause.

In this introduction to the subject I have concentrated on the fundamental problems, on the known methods for solving them in suitably idealized situations, and on the important basic problems still awaiting solution. Recent literature has tended to concentrate on the practical aspects, including the discussion of the properties of specific substances, and it seemed therefore useful to stress the fundamental aspects.

The book is, in the first place, addressed to theoretical physicists, but it should also be useful to experimentalists with a working knowledge of quantum mechanics who want to understand the basis of the models customarily used in dealing with the properties of solids. The book developed from notes which were prepared for a course given in 1953 at the Les Houches summer school in theoretical physics (under the auspices of the University of Grenoble).

The order in which the material is presented was determined by convenience of exposition rather than by the history of the subject. No attempt has been made to give credit to the original authors. Their names can be found in the more detailed review articles and textbooks which are quoted in the bibliography.

Some branches of the theory of solids have developed to such an extent that it was impossible to do them justice within the scope of this small volume. In particular, crystal structures and binding energies, the strength and plasticity of solids, and the electric breakdown of dielectrics are amongst the important topics which have been omitted completely.

I would like to acknowledge my indebtedness to the organizers of the summer school at Les Houches, and in particular to its director, Mrs. Cecile De Witt, for the invitation which provided the occasion to tidy up my ideas about the subject. Discussions with many members of the summer school helped in getting some points stated more clearly. I also received a good deal of constructive criticism from several research students at Birmingham, particularly J. B. Taylor, W. Marshall, and D. A. Greenwood. The latter also very kindly assisted with the proofs. Finally, I would like to express my thanks to the publishers for help in many ways, and to their anonymous proof-reader for many useful suggestions about wording.

R. E. P.

BIRMINGHAM
July 1954

NOTE ON SECOND IMPRESSION

On reprinting this book no major changes have been made, except to correct a few errors, of which the only substantial ones are in equations (7.30), (7.32), (7.34), and (7.40). I am grateful to Professor W. Pauli for drawing my attention to these errors.

Notes containing further explanation, or references to recent work have been added at the end of Chapters I, II, VII, and VIII.

CONTENTS

I. CRYSTAL LATTICES. GENERAL THEORY
1.1 Introduction. Examples of structures — 1
1.2 Dynamical problem. Adiabatic approximation — 4
1.3 Equilibrium — 7
1.4 Types of binding — 9
1.5 Atomic vibrations. Classical mechanics. Linear chain — 11
1.6 Atomic vibrations. Classical mechanics. General solution — 14
1.7 Properties of normal vibrations — 17
1.8 Remark about elastic constants — 22
1.9 Quantum theory — 24

II. CRYSTAL LATTICES. APPLICATIONS
2.1 Specific heat — 27
2.2 Anharmonic terms. Thermal expansion — 31
2.3 Linear term in specific heat — 34
2.4 Thermal conductivity — 40
2.5 Boltzmann equation — 45
2.6 High temperature — 50
2.7 Impurities and size effect — 52

III. INTERACTION OF LIGHT WITH NON-CONDUCTING CRYSTALS
3.1 Statement of problem. Infra-red absorption — 54
3.2 Diffraction of X-rays — 58
3.3 Effect of the atomic vibrations — 63
3.4 Scattering of light — 68
3.5 Scattering of neutrons — 70

IV. ELECTRONS IN A PERFECT LATTICE
4.1 Bloch theorem — 75
4.2 Strong binding — 79
4.3 Nearly free electrons — 83
4.4 Velocity and acceleration — 87
4.5 Many electrons. Statistics — 89
4.6 Specific heat — 93
4.7 Surface problems — 95

V. COHESIVE FORCES IN METALS
5.1 General discussion — 101
5.2 The Wigner–Seitz approximation — 104
5.3 Distorted structures. Linear chain — 108
5.4 Distorted structures. Three dimensions — 112

CONTENTS

VI. TRANSPORT PHENOMENA
6.1 General considerations. Collision time — 115
6.2 Thermal conductivity — 120
6.3 Static obstacles. Impurities and imperfections — 121
6.4 Effect of lattice vibrations. General — 125
6.5 Collisions between electrons — 131
6.6 Collisions at high temperatures — 133
6.7 Low temperatures — 135
6.8 Validity of assumptions — 139

VII. MAGNETIC PROPERTIES OF METALS
7.1 Paramagnetism — 143
7.2 Diamagnetism of free electrons — 144
7.3 Effect of a periodic field — 150
7.4 Hall effect and magneto-resistance — 155

VIII. FERROMAGNETISM
8.1 The Weiss model — 161
8.2 The spin-wave theory. One dimension — 164
8.3 Spin-wave model and ferromagnetism — 170
8.4 The collective electron model — 173
8.5 Neutron scattering — 177
8.6 Remark on magnetization curves — 179
8.7 Anti-ferromagnetism — 181

IX. INTERACTION OF LIGHT WITH ELECTRONS IN SOLIDS
9.1 General outline. Classical theory — 184
9.2 Transitions between bands — 188
9.3 Photoelectric effect — 190
9.4 Non-conducting crystals — 192

X. SEMI-CONDUCTORS AND LUMINESCENCE
10.1 Semi-conductors — 197
10.2 Number of carriers — 198
10.3 Electrical properties — 200
10.4 Density gradients and space charge — 202
10.5 Rectifying contacts — 205
10.6 Electrons not in thermal equilibrium — 207

XI. SUPERCONDUCTIVITY
11.1 Summary of properties — 211
11.2 Outline of Fröhlich–Bardeen theory — 215
11.3 Effect of a magnetic field — 218
11.4 Objections and difficulties — 220

BIBLIOGRAPHY — 222
REFERENCES — 223
LIST OF SYMBOLS — 225
INDEX — 227

I

CRYSTAL LATTICES. GENERAL THEORY

1.1 Introduction. Examples of structures

BY a solid we usually mean a substance which shows some stiffness under shear. Normally such substances have a crystalline structure, and, for the purposes of this book, we shall limit ourselves to crystalline solids. This excludes glasses, and I shall not discuss the question whether glasses should properly be regarded as solids, and which of our results, if any, may be applicable to them.

A crystal lattice may be constructed by repeating a 'unit cell', which may consist of one or more atoms, periodically. The vector leading from a point in one unit cell to the corresponding point in another is called a 'lattice vector' and can be represented as a linear combination with integral coefficients of a small number of basic lattice vectors. The set of lattice vectors determines the 'translation group' of the lattice. All lattices with the same translation group differ by having different unit cells; the simplest of them has just one atom in the unit cell. The translation group is therefore often specified by the name of this simplest lattice belonging to it.

I shall make no attempt to give a complete list of even the more important types of lattices, but I shall give a few examples, which will be used later as illustrations.

We start with cubic lattices, which are defined as remaining unchanged if rotated by 90° about any one of three mutually perpendicular axes. It is then clear that the basic lattice vectors can be stated most simply in relation to these cubic axes.

(a) *Simple cubic lattice.* The unit cell has a single atom, the basic lattice vectors are three vectors of equal length a in the directions of the three cubic axes. In other words, taking Cartesian coordinates along these axes, the basic lattice vectors are

$$(a, 0, 0), \quad (0, a, 0), \quad (0, 0, a),$$

the general lattice vector, and hence the position of any lattice point relative to a given one, being

$$(n_1 a, n_2 a, n_3 a)$$

with arbitrary integers (positive, negative, or zero) n_1, n_2, n_3.

(b) *Body-centred cubic lattice.* One atom per unit cell. Basic lattice vectors

$$(a, 0, 0), \quad (0, a, 0), \quad (0, 0, a), \quad (\tfrac{1}{2}a, \tfrac{1}{2}a, \tfrac{1}{2}a).$$

It is evident that twice the fourth vector equals the sum of the other three, so that the most general lattice point is either

$$(n_1 a, n_2 a, n_3 a)$$

or
$$((n_1+\tfrac{1}{2})a, (n_2+\tfrac{1}{2})a, (n_3+\tfrac{1}{2})a).$$

The first set of points forms a simple cubic lattice, the second consists of all the centres of the cubes formed by adjacent points of the first set, hence the name.

One can look at this lattice as having the translation group of the simple cubic lattice, and two atoms in the unit cell, and sometimes this may be convenient, but by doing so one obscures important relationships, because it looks as if the distance between the two atoms in the unit cell were arbitrary, whereas to preserve the cubic symmetry the lattice must be exactly as above.

(c) *Face-centred cubic lattice.* One atom per unit cell. Basic lattice vectors

$$(a, 0, 0), \quad (0, a, 0), \quad (0, 0, a), \quad (\tfrac{1}{2}a, \tfrac{1}{2}a, 0), \quad (\tfrac{1}{2}a, 0, \tfrac{1}{2}a).$$

The most general lattice point is

$$(n_1 a, n_2 a, n_3 a) \quad \text{or} \quad ((n_1+\tfrac{1}{2})a, (n_2+\tfrac{1}{2})a, n_3 a),$$

or $\quad ((n_1+\tfrac{1}{2})a, n_2 a, (n_3+\tfrac{1}{2})a) \quad \text{or} \quad (n_1 a, (n_2+\tfrac{1}{2})a, (n_3+\tfrac{1}{2})a),$

or, expressed differently,

$$(\tfrac{1}{2}n_1 a, \tfrac{1}{2}n_2 a, \tfrac{1}{2}n_3 a),$$

with n_1, n_2, n_3 integers of which either one or all three are even, i.e. $n_1+n_2+n_3$ must be even. This lattice consists of a simple cubic one and added to it the centres of the faces of each cube. It has the property that, for given distance between neighbouring lattice points, the number of lattice points per unit volume is as large as possible. It is therefore called 'close-packed', because it describes an equilibrium arrangement of hard spheres packed tightly together.

These three lattices cover all the cubic translation groups.

(d) *Simple hexagonal lattice.* One atom per unit cell. The basic lattice vectors are two sides of an equilateral triangle, and a third vector at right angles to the plane of the first two. In Cartesian components

$$(a, 0, 0), \quad (\tfrac{1}{2}a, \tfrac{1}{2}\sqrt{3}a, 0), \quad (0, 0, b).$$

The ratio b/a is not restricted by the symmetry. The general lattice point is, accordingly,

$$((n_1+\tfrac{1}{2}n_2)a, \tfrac{1}{2}\sqrt{3}n_2 a, n_3 b),$$

or, expressed differently,

$$(n_1 . \tfrac{1}{2}a, n_2 . \tfrac{1}{2}\sqrt{3}a, n_3 b),$$

where n_1, n_2, n_3 are integers (positive, negative, or zero), such that n_1 and n_2 are either both even or both odd.

As an important example of a unit cell containing more than one atom, consider the NaCl type lattice. This is a simple cubic lattice in which alternate points are occupied by positive and negative ions (e.g. Na$^+$ and Cl$^-$) respectively. Since these are not identical, the translation group consists only of those displacements which lead from a positive to another positive ion. It is easy to see that this is the translation group of the face-centred cubic lattice, with a spacing equal to twice that of the simple cubic lattice occupied by all the ions together. Hence the unit cell may be taken to consist of

one positive ion at $(0, 0, 0)$,

one negative ion at $(\tfrac{1}{2}a, 0, 0)$,

with the face-centred cubic translation group described above.

This gives the lattice sites

$(\tfrac{1}{2}n_1 a, \tfrac{1}{2}n_2 a, \tfrac{1}{2}n_3 a)$ for positive ions,

$(\tfrac{1}{2}(n_1+1)a, \tfrac{1}{2}n_2 a, \tfrac{1}{2}n_3 a)$ for negative ions,

again with the restriction that $n_1+n_2+n_3$ must be even. The second set can be covered by the same formula as the first, provided we take $n_1+n_2+n_3$ to be odd, and in this form it is evident that the sites of all ions form a simple cubic lattice of spacing $\tfrac{1}{2}a$.

Although in this case the unit cell contains two atoms, their spacing is not arbitrary, but is related to the cubic symmetry. If the positive ions were moved relatively to the negative ones, a structure of much lower symmetry would result.

As a further important example, consider the hexagonal close-packed structure. This is obtained from the simple hexagonal lattice by adding a further plane half-way between the original triangular networks, with lattice sites in the positions corresponding to the centres of one-half of the original triangles.

The unit cell now consists of atoms at

$(0, 0, 0)$ and $(\tfrac{1}{2}a, \tfrac{1}{6}\sqrt{3}a, \tfrac{1}{2}b)$, so that the general lattice point becomes

$$(\tfrac{1}{2}n_1 a,\ \tfrac{1}{2}n_2\sqrt{3}a,\ n_3 b) \quad \text{and} \quad (\tfrac{1}{2}(n_1+1)a,\ \tfrac{1}{2}(n_2+\tfrac{1}{3})\sqrt{3}a,\ (n_3+\tfrac{1}{2})b).$$

The reason for having to use a unit cell of two atoms is that the distance between the two cannot be regarded as a lattice vector, since its repetition does not lead to a site occupied by another atom. As before, any alteration of the distance between the atoms in the unit cell, without changing the translation group, would reduce the symmetry. However, the ratio b/a is still arbitrary. For the particular value

$$b/a = \sqrt{\tfrac{8}{3}} = 1 \cdot 632$$

each atom is surrounded by twelve neighbours at the same distance. This again makes such a pattern suitable for packing hard spheres closely together, the density being, in fact, the same as for the close-packed cubic structure. This property accounts for the name of the lattice, but, unless we are actually dealing with hard spheres, the particular value of b/a is of no significance.

These examples may suffice to illustrate the description of lattices. In general, we have to specify the structure of the unit cell, containing r atoms, by listing their positions,†

$$\mathbf{d}_1,\ \mathbf{d}_2,\ldots,\ \mathbf{d}_r$$

relative to some origin in the unit cell, and to list the lattice vectors $\mathbf{a_n}$, where the suffix $\mathbf{n}$ stands for a set of numbers as in the examples. The general position of a lattice point is then

$$\mathbf{d}_j + \mathbf{a_n}. \tag{1.1}$$

Sometimes it is convenient to choose a unit cell larger than necessary, and then to have all translation vectors equal to integral combinations of three basic vectors, so that the lattice sites are

$$\mathbf{d}_j + n_1 \mathbf{a}_1 + n_2 \mathbf{a}_2 + n_3 \mathbf{a}_3, \tag{1.2}$$

where the n's are arbitrary integers. For instance, in the body-centred cubic lattice described before, $\mathbf{a}_1$, $\mathbf{a}_2$, $\mathbf{a}_3$ are vectors of length a in the direction of the coordinate axes, and

$$\mathbf{d}_1 = (0,\ 0,\ 0), \qquad \mathbf{d}_2 = (\tfrac{1}{2}a,\ \tfrac{1}{2}a,\ \tfrac{1}{2}a).$$

1.2. Dynamical problem. Adiabatic approximation

We next turn to the question of the forces which hold the atoms at or near the sites of a regular crystal structure, and for this we first have to find variables in terms of which the problem can be stated.

† Symbols in bold type indicate vectors or tensors.

§ 1.2 CRYSTAL LATTICES. GENERAL THEORY

The atoms which constitute a solid consist of nuclei and electrons. For a description of the state of the solid it is not, however, necessary to specify the state of all the Z electrons of each atom, since we can eliminate most or all of them by a principle that is familiar from the theory of molecules.† Since the atomic nuclei are much heavier than the electrons, they move much more slowly, and it is therefore reasonable to start from the approximation in which they are taken to be at rest, though not necessarily in the regular positions. Then, if we use **R** as a symbol for all the position vectors of the N nuclei $\mathbf{R}_1, \mathbf{R}_2,..., \mathbf{R}_N$, we may imagine the Schrödinger equation solved for the motion of n electrons, with coordinate vectors $\mathbf{r}_1, \mathbf{r}_2,..., \mathbf{r}_n$, collective symbol **r**, in the field of the nuclei in the configuration **R**. The resulting wave function will be a function of the $3n$ variables **r**, and contain **R** as parameters. The energy eigenvalue will similarly contain **R** as parameters. We may therefore define the lowest energy value $E_0(\mathbf{R})$ and the corresponding eigenfunction $\psi_0(\mathbf{r}, \mathbf{R})$. If we now go over to the real problem in which the nuclei are not held fixed, we may try the assumption that at any time the state of the electrons is described by the same wave function, inserting for **R** the positions of the nuclei at that time. We then merely have to describe the state of motion of the nuclei by a wave function $\phi(\mathbf{R})$, and therefore the wave function of the whole system appears in the form

$$\Psi(\mathbf{r}, \mathbf{R}) = \phi(\mathbf{R}) \cdot \psi_0(\mathbf{r}, \mathbf{R}). \tag{1.3}$$

This is known as the 'adiabatic approximation', since the function $\psi_0(\mathbf{r}, \mathbf{R})$ represents the variation of the electronic state upon adiabatic changes of the parameters.

The condition for (1.3) to represent a good approximation to the solution of the complete Schrödinger equation is usually discussed in the theory of molecules, and one knows that the condition is

$$U\hbar/l \ll \Delta E, \tag{1.4}$$

where U is the velocity of the nuclei, $\hbar$ is Planck's constant divided by 2π, l is the distance by which the nuclei have to move to produce an appreciable change in $\psi_0(\mathbf{r}, \mathbf{R})$, and ΔE is the difference of the first excited electron level, at fixed **R**, from the ground state. One verifies easily that for the inner electrons (e.g. the K shell) this condition is always satisfied.

It may be satisfied for all electrons. This can be the case when the solid is built of chemically saturated units. The simplest such case is that of a solid inert gas such as He, Ne, A,..., when the right-hand side is several electron volts, and the left-hand side considerably less. Another

† See, for example, Slater (1951), Appendix 18.

typical case is that of an ionic lattice like NaCl, in which each of the ions has again a closed-shell configuration. Another case covered by this approximation is that of a molecular solid, like solid H_2, in which each molecule has a saturated electron configuration and a finite excitation energy. A somewhat more complicated example is that of diamond, in which the carbon atoms seem to have homopolar bonds, as in an organic molecule, so that one may again regard the electron state as saturated, but cannot express this in terms of small saturated sub-units.

As an example in which the adiabatic approximation is bound to fail, consider the case of an alkali, like Na. Here each atom is unsaturated, in the sense that it contains a free spin, which is capable of two different orientations with the same energy. N atoms put together will therefore have 2^N states with very similar energies. (The energies will differ somewhat because of the interaction between the atoms, and this is a problem which we shall discuss in detail later on.) If there are that many states within a finite energy interval, their distances are bound to be negligibly small, and the inequality (1.4) cannot hold.

In such cases a complete description of the state of the system must include some electronic variables. It is sufficient, however, to include only the outer electrons in this description. Indeed, once we take away the valency electrons, the remaining ions will form closed shells without degeneracy and with a finite excitation energy. We may therefore, in general, apply the adiabatic approximation to the ions, and imagine the system described in terms of the positions of the ions and of the valency electrons.

In some cases the division between ion core and valency electrons may be ambiguous, or we may be in doubt whether the adiabatic approximation is valid for the last closed shell. We may then always include the electrons of that shell in our description, and we shall see that many qualitative conclusions, at any rate, are not affected by this.

The division I have sketched here is, of course, precisely the division between metals, which contain 'free' electrons, and non-metals, in which all electrons are part of saturated structures. It does not, however, follow that we can predict the nature of a solid immediately from the properties of the atom. For example, the reasoning given above for an alkali would not work if in fact it formed a molecular lattice. Molecular solid Na, in which pairs of atoms form saturated units, exists no doubt in principle, but it is less stable than the ordinary form. To be sure that alkalis are metals we must therefore either fall back on our empirical knowledge that in ordinary conditions they do not form molecular lattices, or attempt a quantitative estimate of the energy of the molecular as compared to the metallic form.

Even in a non-metallic crystal, the adiabatic approximation will be valid only for the electronic state of lowest energy. If we visualize an excited state simply as the excitation of one atom, it is clear that there will be N such states of very similar energy, since any one of the atoms may be excited. Hence the inequality (1.4) is again bound to fail. This problem arises in connexion with the optical spectra of non-metallic crystals, and we shall return to it later.

For the present we shall therefore restrict ourselves to the ground state of a non-metallic crystal, when the adiabatic approximation (1.3) describes the whole system. The function $\phi(\mathbf{R})$ for the motion of the nuclei then satisfies a Schrödinger equation in which the potential consists of the electron energy $E_0(\mathbf{R})$ and of the direct electrostatic interaction between the nuclei. Altogether this will amount to a potential energy $U(\mathbf{R})$.

1.3. Equilibrium

We shall find that, in the problem of the motion of N nuclei under the influence of the potential U, quantum effects are negligible if the temperature is not too low. We therefore begin by discussing the classical problem. Classically the state of lowest energy of the system is obtained by having all nuclei at rest in a configuration which will make $U(\mathbf{R})$ least.

It is to be expected that this most stable configuration will turn out to be one of the lattice structures discussed before. The most stable arrangement is characterized by the following conditions:

(a) It is an equilibrium configuration, i.e. the force on each atom vanishes.

(b) It is in equilibrium against macroscopic displacements, i.e. compression or expansion and shear of the whole lattice (this condition is not a consequence of (a)).

(c) The equilibrium is stable, i.e. the forces will tend to reverse any small displacement, and similarly any change of lattice constant or shear.

(d) All other configurations which satisfy conditions (a), (b), (c) have higher energy.

Of these conditions (a) is automatically satisfied in any structure in which each atom is a centre of symmetry, i.e. in which the reversal of all coordinates, with any given atom as origin, leads to an identical lattice. In that case it follows at once that the change of potential energy caused by displacing an atom by a small amount must equal the change caused by moving it by the same amount in the opposite direction. Hence the change must be an even function of the displacement, and its derivative, which is the force, must vanish in the original position.

All the simple lattices listed above have a centre of symmetry in each atom, except the hexagonal close-packed lattice. In the latter case, there is mirror symmetry about a plane parallel to the first two axes and about three planes, through the third axis and forming angles of 60° with each other. From this symmetry it can again be proved that each atom is in equilibrium.

Similarly, the cubic symmetry implies that the energy must be an even function of any shear, and hence (b) follows for any cubic crystal, except for the change in lattice constant. However, the lattice constant is a variable, and while we cannot hope to prove that a lattice is in equilibrium for an arbitrary value of a, there will, in general, be a particular value of a for which this is true. Evidently this is the value for which the potential energy of the crystal is a minimum. Now the potential energy $U(\mathbf{R})$, regarded as a function of a, tends to $+\infty$ as $a \to 0$, because all atoms repel each other strongly at very close approach. On the other hand, it tends to the energy of isolated atoms (or ions, as the case may be) as $a \to \infty$. As long as the forces are attractive for some value of a, so that, for some a_1, $U(a_1) < U(\infty)$, it is clear that a minimum must exist.

The reasoning for the hexagonal lattice is similar, except that the symmetry would still permit a non-vanishing shear strain, oriented so as to tend to compress the lattice in the plane of the first two axes, and to expand it along the third, or vice versa. This amounts to a change of b/a, and again there will be some value of this ratio for which the potential energy is least.

We see therefore that under very general assumptions about the nature of $U(\mathbf{R})$ there exist lattices of each of the principal symmetry types which satisfy conditions (a) and (b), so that the lattice is in equilibrium. Condition (c), the stability of the equilibrium, requires a better knowledge of the forces, but can be verified in very many cases.

Condition (d), which distinguishes the absolute from any relative minimum of U, is much the hardest. It is in general possible to examine the more obvious alternative structures and compare their energies, but it is hard to rule out the possibility that some other, perhaps less regular, structure might have lower energy still. In fact, from time to time papers appear in which the authors claim to prove that the true equilibrium has a 'mosaic structure', i.e. a periodic distortion of a regular lattice with a period of perhaps a few thousand lattice constants or more. So far all these specific attempts have been disproved and we shall take it for granted that the most stable form of a solid is the perfect crystal lattice, though in practice, of course, it is impossible to attain perfection and, no matter how carefully a crystal is grown, some imperfections will remain.

1.4. Types of binding

A full review of our knowledge of binding energies and stability of lattices would be beyond the scope of this treatment, but, since we shall frequently have to refer to some properties of the potential energy function $U(\mathbf{R})$, it is useful to discuss briefly the most important cases.

The most important types of force are as follows:

(a) *Electrostatic forces.* In an ionic crystal the attraction is mainly due to the Coulomb interaction between point charges. This is particularly amenable to calculation, and a great deal of work has been done on it. The force is a 'two-body' force, i.e. the interaction between two given ions is independent of the positions of any other ions that may be present. It is a central force, i.e. the potential energy describing the interaction between two ions depends only on their distance, so that the force acts along the line joining them. It is not a short-range force, since the variation of the potential (which is proportional to the inverse distance) is rather slow, so that the force on any ion is not predominantly due to its nearest neighbours. For the total binding energy of an ion the effect of distant ions is even more important.

(b) *Van der Waals forces.* This name describes the effect that a neutral and isotropic atom can acquire a polarization under the influence of an electric field, and even two neutral isotropic atoms will induce small dipole moments in each other, due to the fluctuating moments which they possess because of the existence of virtual excited states. As a result, we obtain a potential between an ion and a neutral atom proportional to the inverse fourth power of the distance (the electric field of the ion is proportional to r^{-2}, the induced dipole is proportional to the field, and the energy proportional to the product of the two) and a similar correction term in the interaction of two ions. In the case of two neutral atoms, the van der Waals potential is proportional to the inverse sixth power of their distance. These power laws are valid asymptotically, i.e. for distances large compared with the radii of the atoms or ions. Since they decrease rapidly with distance, they are actually of importance only for small distances, for which the asymptotic law is not valid. For this reason a rigorous description of these forces is difficult, and one is mostly limited to qualitative reasoning.

The van der Waals force is a small correction in the case of ionic crystals, but it is the main attractive force in the case of chemically inert atoms like those of the inert gases, or the main attractive forces between the molecules of a molecular solid like H_2.

It is not precisely a two-body force, since the moment induced in one atom by a second will interact with a third, but in the case of neutral atoms it may be regarded as a two-body force for large separation, when the power laws mentioned above are valid. In the two-body approximation it is also a central force, and it is certainly of short range, so that in general the effects of any but the nearest neighbours are negligible.

(c) *Homopolar binding.* These are forces like those effective in homopolar molecules, and we know they are due to the exchange of electrons between the atoms. In molecular crystals (H_2, Cl_2, etc.) these bonds can easily be localized and we can start from a description of the molecule by the methods of quantum chemistry and then add the relatively weak forces between different molecules. In other cases, however, such as diamond or graphite, each atom shares some valence electrons with each of its neighbours, and it is therefore not possible to single out any given group of atoms that may be regarded as chemically saturated. The quantitative discussion of such forces is not easy. They are probably short-range forces, and, once the bonds have been definitely allocated to pairs of atoms, they may be regarded as two-body forces. They are not, however, always central forces. In a diatomic molecule like H_2 the energy of the chemical bond will depend only on the distance of the two atoms, but, for example, in the graphite or diamond lattice there is a strong tendency for the neighbours of any given carbon atom to arrange themselves in a definite pattern (as there is in an organic molecule) and there is a force resisting the lateral displacement of any of the neighbours.

(d) *Overlap.* If two atoms approach so closely that their electron shells overlap, then there is a strong repulsive force between them. This force is in part electrostatic in nature, since the positive charges of the nuclei are no longer completely shielded from each other, and in part due to the Pauli principle which prevents the electrons from occupying the same part of space without raising their kinetic energy. In this latter respect they are the counterpart of the homopolar forces for the case when no free valency bond is available. These forces may be regarded as central, two-body forces of short range.

§ 1.4   CRYSTAL LATTICES. GENERAL THEORY   11

(e) *Metallic bond*. While we have, for the present, excluded metals from our discussion, it is worth noting that in the case of a metal the presence and motion of the conduction electrons is an important factor in holding the crystal together and in determining its structure. It will be convenient to return to this problem when we know more about conduction electrons, but it is useful to remember that the electrons are free to move over considerable distances, and that their motion is influenced by the arrangement of all atoms they pass, so that the resultant energy is not in the least likely to yield a two-body force, or a central, or short-range, force. This will turn out to be the reason why in metallic elements the stable lattices may be of a very complex kind, quite unlike the examples of high symmetry which were mentioned before.

1.5. Atomic vibrations. Classical mechanics. Linear chain

In practice the atoms even of an ideal crystal lattice will not sit exactly at their equilibrium positions, because of their thermal agitation and for other reasons. We are therefore interested in their motion when near equilibrium. This motion can easily be described for the most general case, but the need for a general notation then tends to obscure the picture, and I shall, therefore, first describe a simple example, the linear chain.

Assume N atoms each of mass M, free to move along a line. It is assumed that the potential energy U is lowest when they are regularly spaced, the distance between any two neighbours being a. Let u_n be the displacement of the nth atom from this equilibrium position. We may then assume U expanded in a Taylor series

$$U - U_0 = \sum_{n,n'} \tfrac{1}{2} A_{n,n'} u_n u_{n'} + \sum_{n,n',n''} \tfrac{1}{6} B_{n,n',n''} u_n u_{n'} u_{n''}, \quad (1.5)$$

where $A_{n,n'} = A_{n',n}$; $B_{n,n',n''} = B_{n',n,n''}$, etc. U_0 is the potential in equilibrium. There are no linear terms since all first derivatives of U must vanish in the equilibrium configuration. The ratio of the coefficients A to the coefficients B is of the dimension of a length, and for reasonable forces this is likely to be of the order of the interatomic distances; the second term on the right-hand side is therefore small if the atomic displacements are small compared with a. This is true in ordinary crystals even at the melting-point, and we can therefore neglect the cubic (and higher) terms, except for certain problems which are entirely dependent on these terms, and to which we shall return.

Neglecting these terms for the present, we find the equations of motion

$$M \ddot{u}_n = - \sum_{n'} A_{n,n'} u_{n'}. \quad (1.6)$$

We note two important properties of the coefficients. Firstly, the force between two atoms which are each displaced by a given amount depends only on their distance apart, i.e. on the difference between their serial numbers n and n'. Hence

$$A_{n,n'} = A(n'-n); \quad A(n'-n) = A(n-n'). \tag{1.7}$$

Secondly, the forces do not change if all atoms are displaced by the same amount in the same direction. This requires

$$\sum_n A(n'-n) = 0. \tag{1.8}$$

We now look for the normal vibrations, i.e. motions in which all atoms vibrate with the same frequency ω radians/sec. We may write

$$u_n(t) = u_n^0 e^{-i\omega t}. \tag{1.9}$$

The complex notation is used here only for mathematical convenience, and we shall, of course, in the end build up real solutions from complex functions of the type (1.9). This gives the equation

$$M\omega^2 u_n^0 - \sum_{n'} A(n'-n) u_{n'}^0 = 0. \tag{1.10}$$

This is a difference equation with constant coefficients for which the solution can be found by a well-known argument.

If we make the substitution

$$n \to n+1,$$

the equation (1.10) remains unchanged. Hence any solution must still be a solution after we have performed the substitution. This new solution may be either essentially identical with the old one, or different. In the first case it can still differ by a constant factor, so that

$$u_{n+1} = e^{ifa} u_n, \tag{1.11}$$

which is general provided we do not for the moment require f to be real. If, on the other hand, the new solution is not proportional to the old one, it follows that the equation has at least two independent solutions, and this freedom can then be used to build linear combinations (which are still solutions) which do satisfy (1.11).

Taking our solutions in this form, we have evidently

$$u_n^0 = \text{constant} \times e^{ifna}, \tag{1.12}$$

and on inserting in (1.10)

$$M\omega^2 = \sum_l A(l) e^{ifla} = \sum_l A(l) \cos fla, \tag{1.13}$$

where in the summation we have written l for $n'-n$, and the last equality follows from the symmetry of the coefficients. We note, firstly, that n

§ 1.5 CRYSTAL LATTICES. GENERAL THEORY

has disappeared from the equation, so that we have reduced the set of equations (1.10) to a single one, thus justifying the form (1.12), and we also see that (1.13) gives us for any f a definite value of ω^2, with

$$\omega^2(f) = \omega^2(-f).$$

We must now consider what are permissible values for f. This depends on what we assume about the ends of the chain. Strictly speaking, our rule (1.7) is valid only for an infinitely long chain, since the forces on atoms near the end are different from those in the middle of the chain. In fact, with realistic assumptions about the forces, even the equilibrium distances will be disturbed near the ends. We can avoid these complications by imagining the atoms arranged on a large circle so that the last atom, $n = N$, is again a distance a from the first, $n = 1$. It is clear that, for large N, the properties of such a cyclic chain will not differ appreciably from those of a chain with open ends, the difference being a specific end effect which can be discussed if necessary. Then our rule (1.7) is rigorously valid, but the displacements must obey the condition

$$u_{n+N} = u_n \quad \text{(cyclic condition)} \tag{1.14}$$

since the serial numbers n and $n+N$ belong to the same atom. (1.14) with (1.12) shows that

$$e^{ifNa} = 1, \tag{1.15}$$

so that f is equal to $2\pi/Na$ times an integer.

On the other hand, since f occurs only in the combination e^{ifna}, nothing is changed if we add to f a multiple of $2\pi/a$, and we can use this to restrict f to the interval

$$-\frac{\pi}{a} < f \leqslant \frac{\pi}{a}. \tag{1.16}$$

There are evidently just N values of f in the interval (1.16) which satisfy (1.15).

Consider now the frequency ω as a function of the 'wave number' f. For the particular value $f = 0$, ω^2 vanishes, as one can see from (1.13) if one remembers (1.8). Since, by (1.13), ω^2 is an even function of f, its expansion for small f will in general start with a term proportional to f^2, so that, for small f, ω is proportional to $|f|$. This means that when the wavelength is large compared with a, the vibrations represent waves with constant wave velocity.

For any f we know that ω^2 is positive, i.e. ω real. Otherwise there would exist a solution in which (1.9) would rise exponentially in time, i.e. in which the displacement from equilibrium would increase indefinitely. This would merely show that the equilibrium was unstable, and that we have not started from the correct stable equilibrium of the system.

A particularly simple case is that in which the force between any but the nearest neighbours is negligible. In that case it is easy to see that $A(l) = 0$, unless $l = 0$ or ± 1, and that, from (1.7) and (1.8),
$$A(1) = A(-1) = -\tfrac{1}{2}A(0),$$
so that
$$M\omega^2 = A(0)(1-\cos fa),$$
$$\omega = (2A(0)/M)^{\frac{1}{2}}|\sin fa/2|.$$

From the solutions thus obtained we can now build up the general solution of the original equation (1.6).

Let
$$u_n = \sum_f q_f e^{ifna}, \qquad (1.17)$$
where the sum is taken over all values of f permitted by (1.15). Then the equation of motion for each of the q_f is
$$\ddot{q}_f + \omega^2(f) q_f = 0, \qquad (1.18)$$
so that they are, indeed, normal coordinates. They will, in general, be complex, and, to ensure that u_n is real, we must have
$$q_{-f} = q_f^*, \qquad (1.19)$$
where the asterisk stands for the complex conjugate. Thus only one-half of the N variables q_f are independent, but, since each of them is complex and contains two real variables, the number of degrees of freedom is just right. (To be precise, (1.19) restricts q_0 and $q_{\pi/a}$ to be real, but they are then independent of other coordinates.)

1.6. Atomic vibrations. Classical mechanics. General solution

We now extend the results of the preceding section to the general case. Let $\mathbf{u}_{n,j}$ be the displacement of the atom which belongs in the jth place of the unit cell which is reached by the lattice vector $\mathbf{a}_n$ from the origin. Then we have, in place of (1.5),
$$U - U_0 = \sum_{j,j'} \sum_{n,n'} \tfrac{1}{2} \mathbf{A}_{j,n;j',n'}\, \mathbf{u}_{j,n}\, \mathbf{u}_{j',n'} +$$
$$+ \sum_{j,j',j''} \sum_{n,n',n''} \tfrac{1}{6} \mathbf{B}_{j,n;j',n';j'',n''}\, \mathbf{u}_{j,n}\, \mathbf{u}_{j',n'}\, \mathbf{u}_{j'',n''}. \qquad (1.20)$$

Here $\mathbf{A}$ and $\mathbf{B}$ are tensors of second and third rank respectively, linking the components of the vectors $\mathbf{u}$ which follow.

It is again true that the $\mathbf{A}$ are symmetric in their indices according to their definition (1.20) and depend only on the relative positions of the relevant unit cells:
$$\mathbf{A}_{j,n;j',n'} = \mathbf{A}_{j,j'}(\mathbf{a}_n - \mathbf{a}_{n'}) = \mathbf{A}_{j',j}(\mathbf{a}_{n'} - \mathbf{a}_n), \qquad (1.21)$$

§ 1.6 CRYSTAL LATTICES. GENERAL THEORY

and again a homogeneous displacement of all atoms in any given direction causes no change:

$$\sum_{\mathbf{n}'}\sum_{j'} \mathbf{A}_{j,j'}(\mathbf{a_n}-\mathbf{a_{n'}}) = 0. \tag{1.22}$$

The equations of motion are

$$M_j \ddot{u}_{j,\mathbf{n}} = -\sum_{j'}\sum_{\mathbf{n}'} \mathbf{A}_{j,j'}(\mathbf{a_n}-\mathbf{a_{n'}})u_{j',\mathbf{n}'}, \tag{1.23}$$

where M_j is the mass of the atom in site j of the unit cell.

We again look for normal vibrations, and we use again the argument that the normal modes either have, or can be chosen to have, the property that upon any translation they multiply by a constant factor. The analogue of (1.9) and (1.12) therefore readily appears in the form

$$\mathbf{u}_{j,\mathbf{n}} = e^{-i\omega t + i\mathbf{f}.\mathbf{a_n}}\mathbf{v}_j, \tag{1.24}$$

where $\mathbf{f}$ is now a vector. $\mathbf{v}_j$ is the displacement of the jth atom in the cell $\mathbf{n} = 0$ at $t = 0$.

Inserting in (1.23) we find

$$M_j \omega^2 \mathbf{v}_j = \sum_{j'} \mathbf{G}_{j,j'}(\mathbf{f})\mathbf{v}_{j'}, \tag{1.25}$$

where
$$\mathbf{G}_{j,j'}(\mathbf{f}) = \sum_{\mathbf{n}} \mathbf{A}_{j,j'}(\mathbf{a_n})e^{i\mathbf{f}.\mathbf{a_n}}. \tag{1.26}$$

The last two equations are the analogue of (1.13). Whereas, however, that equation contained the explicit answer to the problem, we are now still left with (1.25) as a set of $3r$ homogeneous linear equations for the $3r$ displacements of the r atoms in a unit cell. (Remember that $\mathbf{v}$ is a vector and $\mathbf{G}$ a tensor.) These equations have solutions only if the determinant vanishes. This is a function of degree $3r$ in ω^2, and there will in general exist $3r$ roots. It may again be concluded that all these must be real and positive, since otherwise the equilibrium from which we started could not have been stable.

To select the right values for $\mathbf{f}$, we again choose a cyclic condition like (1.14), although this no longer admits any simple physical interpretation. The reason is that this gives us a system with the right number of degrees of freedom, which differs from the real crystal only by surface effects. We assume the crystal to be rectangular, having edges of length L_1, L_2, L_3, respectively, along the three coordinate axes. Then the cyclic condition requires that

$$e^{if_1 L_1} = e^{if_2 L_2} = e^{if_3 L_3} = 1, \tag{1.27}$$

so that the components of $\mathbf{f}$ must be real and multiples of $2\pi/L_1$, $2\pi/L_2$, $2\pi/L_3$, respectively. The actual values of $\mathbf{f}$ are of little interest, since their exact position is linked to the cyclic boundary condition, but the important consequence of (1.27) is that the permitted values of **f** are uniformly distributed over f space, and that their density is $L_1 L_2 L_3/(2\pi)^3$, i.e. $V/(2\pi)^3$, where V is the volume.

Next we require the analogue of (1.16). **f** occurs only in the combination $e^{i\mathbf{f}.\mathbf{a_n}}$, where $\mathbf{a_n}$ is a lattice vector. Hence two vectors **f** and **f'** describe the same normal mode if
$$(\mathbf{f'}-\mathbf{f}).\mathbf{a_n}$$
is a multiple of 2π for any **n**.

We introduce a set of vectors **K** by the requirement that
$$e^{i\mathbf{K}.\mathbf{a_n}} = 1 \text{ for any } \mathbf{n}, \tag{1.28}$$
and these again form a lattice (evidently the sum of two vectors satisfying (1.28) will also satisfy it). This is known as the 'reciprocal lattice'. It depends only on the translation group of the original lattice, not on the structure of the unit cell.

For a simple cubic lattice, the reciprocal lattice is also simply cubic, with lattice constant $2\pi/a$, as one sees at once from (1.28) by inserting the three basic translation vectors.

As another example, consider the cubic body-centred lattice. Here
$$\mathbf{a_n} = (\tfrac{1}{2}n_1 a, \tfrac{1}{2}n_2 a, \tfrac{1}{2}n_3 a)$$
with n_1, n_2, n_3 either all even or all odd. (1.28) then means that
$$K_1 n_1 + K_2 n_2 + K_3 n_3 = 4\pi/a \text{ times an integer,}$$
for any permissible set of n's. Choosing for the n's the particular cases $(2, 0, 0)$, $(0, 2, 0)$, and $(0, 0, 2)$, we see that each of the components of **K** must be an integral multiple of $2\pi/a$. In addition, the case $(1, 1, 1)$ shows that $K_1+K_2+K_3$ is an even multiple of $2\pi/a$. We have the final result
$$\mathbf{K} = \frac{2\pi}{a}(\kappa_1, \kappa_2, \kappa_3), \tag{1.29}$$
where $\kappa_1, \kappa_2, \kappa_3$ are arbitrary integers whose sum is even. Thus the reciprocal to a cubic body-centred lattice of spacing a is a face-centred cubic lattice of spacing $4\pi/a$. Conversely the reciprocal of a face-centred is a body-centred cubic lattice.

Returning to the definition of **f**, we see now that two vectors are equivalent if they differ by a lattice vector of the reciprocal lattice. Now we choose, of all these equivalent vectors, the one of the smallest magnitude. The values of **f** so selected will cover a region in **f** space defined as follows: Draw in the reciprocal lattice the lines from the origin to all other lattice points in the neighbourhood. Draw the plane which bisects each of these lines, and discard that half-space bounded by the plane which does not contain the origin. This will finally leave a polyhedron which we call the basic cell of the reciprocal lattice, and which contains all our selected **f** values.

In a simple cubic lattice, this basic cell is a cube of edge $2\pi/a$; in the body-centred case it is a regular dodecahedron.

To sum up, the normal vibrations of a lattice can be expressed as progressive waves of the form (1.24) where **f** is a vector within the basic cell of the reciprocal lattice. The density of permitted values for **f** is $V/(2\pi)^3$. For each given **f**, there are $3r$ different modes, corresponding to different solutions of (1.25), in general with different frequencies ω. We label these solutions $\mathbf{v}_j(\mathbf{f}, s)$ and the frequencies $\omega(\mathbf{f}, s)$, where $s = 1, 2, ..., 3r$.

The most general motion of the atoms can again be expressed as a superposition (cf. 1.17)

$$\mathbf{u}_{j,\mathbf{n}} = \sum_{\mathbf{f},s} q_{\mathbf{f},s}(t) e^{i\mathbf{f}.\mathbf{a}_\mathbf{n}} \mathbf{v}_j(\mathbf{f}, s), \qquad (1.30)$$

where the normal coordinates q satisfy the equation

$$\ddot{q}_{\mathbf{f},s} + \{\omega(\mathbf{f}, s)\}^2 q_{\mathbf{f},s} = 0. \qquad (1.31)$$

For some purpose it is convenient to separate $q_{\mathbf{f},s}$ into the part which has the time-dependence of (1.24) with positive frequency, and the part with negative frequency. This amounts to separating the elastic waves into those travelling in opposite directions. We write

$$q_{\mathbf{f},s} = Q(\mathbf{f}, s) + Q(\mathbf{f}, -s), \qquad (1.32)$$

where
$$\dot{Q}(\mathbf{f}, s) = -i\omega(\mathbf{f}, s) Q(\mathbf{f}, s) \qquad (1.33)$$

with the convention that $\omega(\mathbf{f}, s)$ is negative when s is negative. Conversely,

$$Q(\mathbf{f}, s) = -\frac{1}{2i\omega(\mathbf{f}, s)} \{\dot{q}(\mathbf{f}, s) - i\omega(\mathbf{f}, s) q(\mathbf{f}, s)\}. \qquad (1.34)$$

This last equation holds for both positive and negative s. The $6Nr$ variables Q may be used instead of the $3Nr$ variables q and their time derivatives. To avoid ambiguity we shall use the label σ in place of s in all sums covering both positive and negative values so that

$$\left. \begin{array}{l} s = 1, 2, ..., 3r \\ \sigma = \pm 1, \pm 2, ..., \pm 3r \end{array} \right\}. \qquad (1.35)$$

1.7. Properties of normal vibrations

It is convenient to collect a number of mathematical and physical relations that will be needed. Firstly it is evident from (1.26) that

$$\mathbf{G}_{j,j'}(-\mathbf{f}) = \mathbf{G}^*_{j,j'}(\mathbf{f}),$$

so that the coefficients of (1.25) change to their complex conjugates upon reversal of **f**. Hence the solution for $-\mathbf{f}$ can be obtained in the form

$$\mathbf{v}_j(-\mathbf{f}, s) = \mathbf{v}_j^*(\mathbf{f}, s), \qquad (1.36)$$

and the reality condition for the atomic displacements **u**, according to (1.30), is

$$q_{-\mathbf{f},s} = q_{\mathbf{f},s}^* \qquad (1.37)$$

as in (1.19). Hence from (1.34) also

$$Q(-\mathbf{f}, -\sigma) = Q^*(\mathbf{f}, \sigma). \qquad (1.38)$$

Different normal vibrations are orthogonal to each other in the following sense:

$$\sum_j \sum_\mathbf{n} M_j \{\mathbf{v}_j(\mathbf{f}, s) e^{i\mathbf{f} \cdot \mathbf{a}_\mathbf{n}}\}^* \cdot \{\mathbf{v}_j(\mathbf{f}', s') e^{i\mathbf{f}' \cdot \mathbf{a}_\mathbf{n}}\} = 0 \qquad (1.39)$$

unless $\mathbf{f} = \mathbf{f}'$, $s = s'$.

To prove this, assume first that $\mathbf{f} \neq \mathbf{f}'$. Then the summation over **n** yields the factor

$$S = \sum_\mathbf{n} e^{i(\mathbf{f}'-\mathbf{f}) \cdot \mathbf{a}_\mathbf{n}}. \qquad (1.40)$$

That this sum vanishes can be seen in the following way. The summation is extended over the whole lattice of the crystal. Now suppose each lattice point of the crystal displaced by a particular lattice vector, say **a**. This changes every lattice point into another lattice point (by the definition of lattice vectors) and therefore merely results in a re-labelling of the terms of the sum. Thus the sum does not change. On the other hand, each term in the sum is multiplied by the factor $e^{i(\mathbf{f}'-\mathbf{f}) \cdot \mathbf{a}}$, and hence the whole sum is multiplied by this factor on displacement. Hence

$$S = S e^{i(\mathbf{f}'-\mathbf{f}) \cdot \mathbf{a}}.$$

This means that either $S = 0$ or $(\mathbf{f}'-\mathbf{f}) \cdot \mathbf{a}$ is a multiple of 2π as long as **a** is any lattice vector. Using the definition (1.28), this shows that S vanishes unless $\mathbf{f}'-\mathbf{f}$ is a vector of the reciprocal lattice, and therefore $\mathbf{f}'$ and $\mathbf{f}$ are equivalent.

It remains to discuss (1.39) when $\mathbf{f} = \mathbf{f}'$. Then the sum over **n** yields the number N of unit cells in the crystal, and (1.39) becomes

$$N \sum_j M_j \mathbf{v}_j^*(\mathbf{f}, s) \cdot \mathbf{v}_j(\mathbf{f}', s).$$

Now it is easy to prove that this sum vanishes if $\omega(\mathbf{f}, s) \neq \omega(\mathbf{f}', s')$, because, if we multiply the conjugate complex equation to (1.25) by $\mathbf{v}_j(\mathbf{f}', s')$ and sum over j, then subtract the corresponding quantity with $\mathbf{f}, s, i$ interchanged with $\mathbf{f}', s', -i$, the right-hand side cancels by virtue of (1.21) and the left-hand side is the required sum multiplied by $\omega^2 - \omega'^2$.

§1.7 CRYSTAL LATTICES. GENERAL THEORY

If more than one normal vibration with a given $\mathbf{f}$ have the same frequency, then the $\mathbf{v}_j(\mathbf{f}, s)$ are not uniquely determined, since the set of linear homogeneous equations (1.25) then has several independent solutions. In that case we may, as usual, choose our basic set of solutions in such a way as to satisfy (1.39).

The value of the sum (1.39) when $\mathbf{f} = \mathbf{f}'$, $s = s'$ is at our choice since the defining equation (1.25) leaves a constant factor free in $\mathbf{v}_j$. We choose for the normalization of the solutions the convention

$$\sum_j M_j |\mathbf{v}_j(\mathbf{f}, s)|^2 = M^c,$$

where M^c is the mass of a unit cell. Then we can express the orthogonality and normalization in one equation

$$\sum_j \sum_\mathbf{n} M_j \{\mathbf{v}_j(\mathbf{f}, s) e^{i\mathbf{f}.\mathbf{a}_\mathbf{n}}\}^* \cdot \{\mathbf{v}_j(\mathbf{f}', s') e^{i\mathbf{f}'.\mathbf{a}_\mathbf{n}}\} = M^{(N)} \delta_{\mathbf{ff}'} \delta_{ss'}, \qquad (1.41)$$

where $M^{(N)} = NM^c$ is the mass of the whole crystal of N unit cells, and $\delta_{\mathbf{ff}'}$ is the usual Kronecker symbol, i.e. is unity when $\mathbf{f} = \mathbf{f}'$ and zero otherwise.

The content of equations like (1.41) tends to get obscured by the number of variables, and it pays, therefore, to use a shorthand notation where possible. For this purpose we use a symbol ν to stand for $\mathbf{n}, j$, and the three space directions, and ϕ to stand for $\mathbf{f}$ and s. Then if we let $v_\nu(\phi)$ be one of the components of

$$\mathbf{v}_j(\mathbf{f}, s) e^{i\mathbf{f}.\mathbf{a}_\mathbf{n}},$$

we can write (1.41) as

$$\sum_\nu M_\nu v_\nu^*(\phi) v_\nu(\phi') = M^{(N)} \delta_{\phi\phi'}. \qquad (1.42)$$

In the same way we can write the most general disturbance as

$$u_\nu = \sum_\phi q(\phi) v_\nu(\phi), \qquad (1.43)$$

and it follows from this and (1.42) in the usual way that

$$q(\phi) = \frac{1}{M^{(N)}} \sum_\nu M_\nu v_\nu^*(\phi) u_\nu. \qquad (1.44)$$

If we insert (1.44) in (1.43), we find

$$u_\nu = \sum_\phi v_\nu(\phi) \sum_{\nu'} \frac{M_{\nu'}}{M^{(N)}} v_{\nu'}^*(\phi) u_{\nu'}.$$

Since, for an arbitrary disturbance, the right-hand side must be identical with the left-hand side, we find the 'completeness relation'

$$\sum_\phi v_\nu(\phi) v_{\nu'}^*(\phi) = \frac{M^{(N)}}{M_\nu} \delta_{\nu\nu'}, \qquad (1.45)$$

which is also known to be a direct algebraic consequence of (1.42) provided the number of values of ϕ is the same as that for ν. (In our case the number of values of either is $3rN$.)

We record, for future reference, the expression for the total kinetic energy,
$$T = \tfrac{1}{2}\sum_\nu M_\nu \dot{u}_\nu^2 = \tfrac{1}{2}\sum_\nu M_\nu \sum_{\phi,\phi'} v_\nu^*(\phi)v_\nu(\phi')\dot{q}^*(\phi)\dot{q}(\phi'),$$
which, by (1.43), becomes
$$T = \tfrac{1}{2}M^{(N)}\sum_\phi |\dot{q}(\phi)|^2 = \tfrac{1}{2}M^{(N)}\sum_{\mathbf{f},s}|\dot{q}(\mathbf{f},s)|^2, \tag{1.46}$$
where in the last form we have reverted to the explicit notation. We could similarly transform the potential energy, but its form is at once evident from the fact that the equation of motion must be (1.31), which requires
$$U - U_0 = \tfrac{1}{2}M^{(N)}\sum_{\mathbf{f},s}\{\omega(\mathbf{f},s)\}^2|q(\mathbf{f},s)|^2. \tag{1.47}$$

Lastly we can introduce the variables Q by (1.32) and (1.33) and find for the total energy
$$T + U - U_0 = M^{(N)}\sum_{\mathbf{f}}\sum_s\{|Q(\mathbf{f},s)|^2 + |Q(\mathbf{f},-s)|^2\}\{\omega(\mathbf{f},s)\}^2$$
$$= M^{(N)}\sum_{\mathbf{f}}\sum_\sigma |Q(\mathbf{f},\sigma)|^2\{\omega(\mathbf{f},\sigma)\}^2. \tag{1.48}$$
In the last form the sum extends over positive and negative σ, by the convention (1.35).

The actual solution of (1.25) and the determination of the normal modes and frequencies is very laborious even if the force coefficients **A** or **G** are known, and even for $r = 1$ it requires the solution of three simultaneous equations for each $\mathbf{f}$, corresponding to the three components of $\mathbf{v}$. However, for special cases the answer may be obvious, for example when the direction of $\mathbf{f}$ has a simple relation to the axes of symmetry of the lattice; e.g. in a cubic lattice, if $\mathbf{f}$ is in the direction of one of the main axes, the solutions of (1.25) must have $\mathbf{v}$ parallel either to the same, or to one of the two other axes, the last two cases giving the same frequency. Similarly, if $\mathbf{f}$ has the direction of the space diagonal of the cube, $\mathbf{v}$ is either parallel or perpendicular to $\mathbf{f}$; in the latter case the frequency does not depend on its orientation. If $\mathbf{f}$ is in the direction of a face diagonal, $\mathbf{v}$ is still parallel or perpendicular to it, but in that case the two directions perpendicular to $\mathbf{f}$ and either parallel or normal to the respective cube face have different frequencies.

Another case in which the solution is easy is that of infinite wavelength, $\mathbf{f} = 0$. We note that, for $\mathbf{f} = 0$, the equations (1.25) have solutions for

§ 1.7   CRYSTAL LATTICES. GENERAL THEORY   21

zero frequency. These solutions consist of a uniform displacement of all atoms in the same direction, so that the vector $\mathbf{v}_j$ is the same for all j. Then the right-hand side of (1.25) vanishes, because as a consequence of (1.22) and (1.26)

$$\sum_{j'} \mathbf{G}_{j,j'}(0) = 0.$$

Since the direction of $\mathbf{v}$ is still arbitrary, there are three independent solutions of this kind. They are the only ones with $\mathbf{f} = 0$ if we have a simple lattice with $r = 1$. Otherwise there are still $3r-3$ other solutions of higher frequency. These represent vibrations in which the displacement of corresponding atoms in different unit cells is the same, but the atoms in the same cell move relatively to each other. It is easy to see from the orthogonality to the zero-frequency vibrations that, in the case of $r = 2$ (e.g. the alkali halides) the two atoms in a unit cell must move in opposite directions.

Now consider the case of a small, but finite, $\mathbf{f}$. The solutions will now lie close to those for $\mathbf{f} = 0$,† in other words, there are always three modes of vibration in which the frequency is small; though for $r > 1$ there are also $3r-3$ modes of high frequency. The low-frequency vibrations for small f represent sound waves in the crystal. Since for $f = 0$ the displacements of all atoms in the unit cell are equal, it is evident that for small f they will still be nearly equal. The three different solutions for a given small $\mathbf{f}$ will therefore differ by the direction of this common displacement. We can therefore speak of sound waves of three different polarizations.

We know that in an isotropic medium sound waves can be longitudinal or transverse, and that for a given direction of propagation the two transverse waves have the same velocity, which, however, is less than that of the longitudinal wave.

In the anisotropic crystal the three directions are in general neither longitudinal nor transverse, and they correspond to three different velocities. For special cases the situation simplifies. For example, if in a cubic crystal $\mathbf{f}$ is in the direction of a main axis, the symmetry of the structure requires that the solutions of (1.25) for small $\mathbf{f}$ approximate to one longitudinal and two transverse positions, and that the frequencies of the last two are equal.

Let us assume we expand the values of ω^2 for which (1.25) has a solution in powers of $\mathbf{f}$; the constant term vanishes since we have already seen that for $\mathbf{f} = 0$ also $\omega^2 = 0$. The linear term vanishes since, by virtue

† Except in the case of the 'optical' branch of ionic crystals, see § 3.1.

of (1.32), **f** and $-$**f** must belong to the same frequency. Hence the leading term in ω^2 will be proportional to f^2, and for any fixed direction and fixed polarization the frequency ω is proportional to the wave vector **f**, and their ratio, the velocity of sound, is independent of the wavelength, though, of course, dependent on direction and polarization. From the way this result was obtained, it clearly applies only in the limit of long waves.

This is the region in which the theory of elasticity is applicable, and in fact knowledge of the sound velocity is equivalent to knowledge of the elastic constants.

1.8. Remark about elastic constants

A discussion of the problem of elasticity should, logically speaking, have preceded the work on vibrations, but it was more convenient to introduce our notation through the general vibration problem.

The general force constants **A**, whose knowledge we have assumed in (1.20), must define the elastic constants of the crystal. An elastic deformation is one in which the displacement varies slowly from cell to cell, so that the $\mathbf{u}_{j,\mathbf{n}}$ can be regarded locally as slowly varying functions of **n**. Their dependence on j, however, will in general be complicated; as the crystal is deformed, the different atoms in a unit cell will move by different amounts. In general, this means that one still has to solve a set of simultaneous equations of the type (1.25), and we shall not discuss this problem in detail.

The situation is simple if we are dealing either with a simple lattice ($r = 1$), or with a lattice (such as the NaCl type) in which the relative position of the two particles in the unit cell is determined by symmetry considerations even when the crystal is elastically deformed. I shall elaborate only the case $r = 1$.

First, let us understand the procedure by looking at the case of the linear chain. We therefore evaluate the quadratic term of (1.5) which, remembering (1.7), may be written

$$U - U_0 = \tfrac{1}{2} \sum_{n,l} A(l) u_n u_{n+l}. \tag{1.49}$$

Here we may regard u_{n+l} as a continuous function of l, which can be expanded in a Taylor series:

$$U - U_0 = \tfrac{1}{2} \sum_n \sum_l A(l) u_n \left(u_n + \frac{du}{dn} l + \frac{1}{2} \frac{d^2 u}{dn^2} l^2 + \ldots \right). \tag{1.50}$$

We now carry out the summation term by term. The first term in the

§ 1.8   CRYSTAL LATTICES. GENERAL THEORY

bracket contributes nothing because of (1.8). The second contributes nothing because, by (1.7), $A(l)$ is an even function of l. If we define

$$\tfrac{1}{2} \sum_l A(l) l^2 = -E/a, \tag{1.51}$$

we are left with

$$U - U_0 = -\frac{1}{2} \sum_n \frac{E}{a} u_n \frac{d^2 u}{dn^2}. \tag{1.52}$$

Here the sum contains only quantities which vary slowly, and we may replace the summation by an integration,

$$U - U_0 = -\tfrac{1}{2} E \int u \frac{d^2 u}{dx^2} dx = \tfrac{1}{2} E \int \left(\frac{du}{dx}\right)^2 dx. \tag{1.53}$$

The boundary term vanishes because of the cyclic condition (1.14).

Equation (1.53) is the right expression for the elastic energy of a linear chain, and E is its elastic constant. For the particular case of interaction between nearest neighbours only, (1.51) gives

$$E = aA(1) = \tfrac{1}{2} a A(0).$$

Since the mass per unit length is M/a, the velocity of sound waves is given by

$$c^2 = \frac{a^2 A(0)}{2M},$$

which evidently makes c agree with the ratio ω/f for small f, as previously calculated.

For the three-dimensional case with one atom per unit cell, we can apply the same transformation to (1.20), with the result

$$U - U_0 = \tfrac{1}{2} \sum_{\mu\nu\sigma\tau} E_{\mu\nu\sigma\tau} \int \left(\frac{\partial u_\sigma}{\partial x_\mu} \frac{\partial u_\tau}{\partial x_\nu}\right) d^3 x, \tag{1.54}$$

where μ, ν, σ, τ each take the values 1, 2, 3 and refer to the three space coordinates. The differential d^3x is the volume element in three dimensions, and $E_{\mu\nu\sigma\tau}$ is an elasticity coefficient defined as

$$E_{\mu\nu\sigma\tau} = \frac{1}{2v_0} \sum_{l} a_\mu(l) a_\nu(l) A_{\sigma\tau}(\mathbf{a}_l), \tag{1.55}$$

where $a_\mu(l)$, $a_\nu(l)$ are the μ and ν components of the lattice vector $\mathbf{a}_l$, and $A_{\sigma\tau}$ is the σ, τ component of the tensor $\mathbf{A}$. v_0 is the volume of a unit cell.

We have not yet made use of the fact that the total potential energy will evidently be unchanged if the crystal is rotated without change of form, and this implies a symmetry of E between μ and σ, and between ν and τ. Hence (1.54) may also be written

$$\frac{1}{2}\sum_{\mu,\nu}\sum_{\sigma,\tau} E_{\mu\nu\sigma\tau}\left\{\frac{1}{2}\left(\frac{\partial u_\sigma}{\partial x_\mu}+\frac{\partial u_\mu}{\partial x_\tau}\right)\frac{1}{2}\left(\frac{\partial u_\tau}{\partial x_\nu}+\frac{\partial u_\nu}{\partial x_\tau}\right)\right\}, \tag{1.56}$$

which is the more familiar form expressing the elastic energy as a quadratic function of the strain components.

The symmetry of the crystal gives rise to many identities between these coefficients, and it is well known that in the case of a cubic crystal only three of these coefficients are independent.

A case of particular interest is that of uniform compression, for which

$$\frac{\partial u_\sigma}{\partial x_\mu}=\frac{\epsilon}{3}\delta_{\mu\sigma},$$

where $\delta_{\mu\sigma}$ is the usual Kronecker symbol and ϵ is the relative increase in volume. The corresponding energy increase per unit volume is then, from (1.54),

$$\frac{1}{2}\frac{\epsilon^2}{9}\sum_{\mu,\nu} E_{\mu\nu\mu\nu}. \tag{1.57}$$

We can therefore find the bulk modulus (inverse compressibility)

$$k=\tfrac{1}{9}\sum_{\mu,\nu} E_{\mu\nu\mu\nu}=\tfrac{1}{18}\sum_{\mathbf{l}}\sum_{\mu,\nu} a_\mu(\mathbf{l})A_{\mu\nu}a_\nu(\mathbf{l}). \tag{1.58}$$

1.9. Quantum theory

Before applying our results to practical problems, it is convenient to derive the corresponding equations in quantum mechanics, so that we can then obtain for each problem both the classical and the quantum result.

As in all cases concerned with harmonic motion, the extension to quantum mechanics is straightforward, since the important equations (1.31) and (1.30) are still valid, except that the variables q and u are now to be regarded as operators. The $\mathbf{u}_{j,\mathbf{n}}$, being coordinates of a particle of mass M_j, satisfy the commutation law,

$$[\dot{u}_\sigma(j,\mathbf{n}),u_\tau(j',\mathbf{n}')]=-\frac{i\hbar}{M_j}\delta_{\sigma\tau}\delta_{\mathbf{n}\mathbf{n}'}\delta_{jj'}, \tag{1.59}$$

where σ and τ again specify the space directions. In addition every u commutes with every other u, and every $\dot{u}$ with every other $\dot{u}$.

Using (1.44),

$$[\dot{q}^*(\phi),q(\phi')]=\frac{1}{(M^{(N)})^2}\sum_{\nu,\nu'} M_\nu M_{\nu'} v_\nu(\phi) v_\nu^*(\phi')[\dot{u}_\nu,u_\nu],$$

§ 1.9   CRYSTAL LATTICES. GENERAL THEORY

which from (1.59) is

$$-i\hbar \sum \frac{M_\nu}{(M^{(N)})^2} v_\nu(\phi) v_\nu^*(\phi'),$$

and with the orthogonality relation (1.42) this becomes

$$[\dot{q}^*(\phi), q(\phi')] = -\frac{i\hbar}{M^{(N)}} \delta_{\phi\phi'}, \tag{1.60}$$

which shows that $q(\phi)$ and $M^{(N)}\dot{q}^*(\phi)$ are canonically conjugate variables. This gives us the commutation laws of the Q introduced in (1.34):

$$[Q(\mathbf{f},\sigma), Q^*(\mathbf{f}',\sigma')] = \frac{\hbar}{2M^{(N)}\omega(\mathbf{f},\sigma)} \delta_{\mathbf{f}\mathbf{f}'} \delta_{\sigma\sigma'}, \tag{1.61}$$

which, with the equation of motion

$$\dot{Q}(\mathbf{f},\sigma) = -i\omega(\mathbf{f},\sigma) Q(\mathbf{f},\sigma), \tag{1.62}$$

determines the matrices of the Q.

These can be obtained at once from the remark that the relations (1.61) and (1.62) for Q and Q^* are identical, except for scale factors, with those for the quantities $\xi = \frac{1}{2}(x + (1/i\omega)\dot{x})$ and ξ^*, where x is the coordinate of a simple harmonic oscillator. By analogy we find that the system is described by a set of quantum numbers $N(\mathbf{f}, s)$, defined for positive s, which take non-negative integral values; $Q(\mathbf{f}, s)$ has a non-vanishing matrix element only from a state with quantum number $N(\mathbf{f}, s)$ to that in which $N(\mathbf{f}, s)$ has increased by 1, while all other quantum numbers remain unchanged. This matrix element is

$$(N|Q(\mathbf{f},s)|N+1) = (N+1|Q^*(\mathbf{f},s)|N) = (\hbar/2M^{(N)}\omega(\mathbf{f},s))^{\frac{1}{2}}\sqrt{(N+1)}. \tag{1.63}$$

Because of the reality condition, the second of these quantities is also the matrix for $Q(-\mathbf{f}, -s)$ and we can avoid the need for restricting s to positive values if we also define quantum numbers for negative s by

$$N(-\mathbf{f}, -\sigma) = -N(\mathbf{f}, \sigma) - 1. \tag{1.64}$$

With this definition (1.63) holds for both negative and positive values of σ.

The energy of the crystal, which is, of course, the sum of kinetic and potential energy, takes the form

$$E = \sum_\mathbf{f} \sum_s \{N(\mathbf{f},s) + \tfrac{1}{2}\}\hbar\omega(\mathbf{f},s) + U_0, \tag{1.65}$$

showing the familiar oscillator form. The term $\frac{1}{2}$ in the bracket represents the 'zero-point' energy and is due to the fact that even in the state of lowest energy the atoms cannot be exactly in their equilibrium positions, since an exact localization would require a large uncertainty in their velocities, and hence a large kinetic energy.

Instead of saying that the $(\mathbf{f}, s)$ oscillator is excited to its $N(\mathbf{f}, s)$th state, one may also say that there are $N(\mathbf{f}, s)$ quanta of vibration with wave number $\mathbf{f}$ and polarization s, and such quanta, which have the same relation to sound waves as light quanta to light waves, are called 'phonons'.

Using the somewhat formal definition (1.64), we may write the energy also in the more symmetrical form

$$E = \tfrac{1}{2} \sum_{\mathbf{f}} \sum_{s} N(\mathbf{f}, \sigma) \hbar \omega(\mathbf{f}, \sigma), \tag{1.66}$$

where now σ takes both positive and negative values; but the simplicity of this form is deceptive, since it tends to obscure the fact that of the quantum numbers occurring in it only one-half are independent.

Note added January 1956

It should be emphasized that some of the formulae of this chapter are valid only for lattices in which each atom forms a centre of symmetry. Otherwise the second equality in equation (1.21) does not necessarily hold. The force constants remain unchanged only if in addition to exchanging the two atoms concerned, one also interchanges the appropriate space directions, in other words the components of the tensor **A**, which in the general case are not symmetrical.

II
CRYSTAL LATTICES. APPLICATIONS

2.1. Specific heat

If the crystal is in thermal equilibrium at temperature T, it is well known that the probability of a quantum state of energy E is $e^{-\beta E}$, where

$$\beta = \frac{1}{kT}, \qquad (2.1)$$

k being Boltzmann's constant. With the result (1.65) it follows then that the mean energy of each oscillator is (using a bar over a quantity to indicate a statistical average)

$$\overline{E(\mathbf{f},s)} = \hbar\omega(\mathbf{f},s)\{\overline{N(\mathbf{f},s)}+\tfrac{1}{2}\} = \frac{\sum_{N=0}^{\infty}(N+\tfrac{1}{2})\hbar\omega e^{-\beta\hbar\omega(N+\frac{1}{2})}}{\sum_{N=0}^{\infty}e^{-\beta\hbar\omega(N+\frac{1}{2})}}$$

$$= -\frac{\partial}{\partial\beta}\log\sum e^{-\beta\hbar\omega(N+\frac{1}{2})} = \tfrac{1}{2}\hbar\omega + \frac{\hbar\omega}{e^{\beta\hbar\omega}-1} = \tfrac{1}{2}\hbar\omega\coth(\tfrac{1}{2}\beta\hbar\omega). \qquad (2.2)$$

The energy of the whole crystal is

$$E = U_0 + E_Z + E_T,$$

where E_Z is the zero-point energy

$$E_Z = \sum_{\mathbf{f},s} \tfrac{1}{2}\hbar\omega(\mathbf{f},s), \qquad (2.3)$$

and E_T is the thermal energy

$$E_T = \sum_{\mathbf{f},s} \overline{N(\mathbf{f},s)}\hbar\omega(\mathbf{f},s) = \sum_{\mathbf{f},s}\frac{\hbar\omega(\mathbf{f},s)}{e^{\beta\hbar\omega(\mathbf{f},s)}-1}. \qquad (2.4)$$

The evaluation of the sum (2.4) is, in general, very difficult, and it requires a knowledge of the frequency as a function of $\mathbf{f}$ and s.

However, the calculation can be carried further in the two limits of high and low temperatures. Consider first a temperature so high that for all vibrations

$$\hbar\omega(\mathbf{f},s) \ll kT.$$

This means that the summand in (2.4) can be expanded in a power series, and the three leading terms give

$$E_T = \sum_{\mathbf{f},s}\frac{1}{\beta}[1-\tfrac{1}{2}\beta\hbar\omega(\mathbf{f},s)+\tfrac{1}{12}\{\beta\hbar\omega(\mathbf{f},s)\}^2-\ldots]$$

$$= \sum_{\mathbf{f},s}\left[kT - \tfrac{1}{2}\hbar\omega(\mathbf{f},s) + \frac{1}{12}\frac{1}{kT}\{\hbar\omega(\mathbf{f},s)\}^2 - \ldots\right]. \qquad (2.5)$$

The first term amounts to $3rNkT$, which is precisely the classical result for a system of $3rN$ vibrational degrees of freedom. The second term just cancels the zero-point energy, so that at high temperatures not merely the specific heat, but also the energy content approaches asymptotically the classical result. The third term, which vanishes at infinite T, and which measures the deviation from the classical result, is seen to be of the second order in the quantum constant $\hbar$. This is a very common feature of quantum statistics. For estimating this term one must know something of the vibration spectrum. However, it is not necessary to solve the equations (1.25) for this purpose, since the sum of the squares of the frequencies can be obtained directly in terms of the coefficients of the equation.

One can, indeed, prove easily from the orthogonality relations of § 1.7 and from (1.25) that

$$\sum_{\mathbf{f},s} \{\omega(\mathbf{f},s)\}^2 = N \sum_j \frac{1}{M_j} \langle \mathbf{G}_{jj} \rangle = N \sum_j \frac{1}{M_j} \langle \mathbf{A}_{jj}(0) \rangle, \qquad (2.6)$$

where the symbol $\langle ... \rangle$ indicates the diagonal sum or 'trace' of the tensor. The last sum has a fairly direct interpretation: it represents the sum of the potential energies obtained by displacing in turn each atom of the crystal in each coordinate direction by an infinitesimal amount, dividing by the square of the displacement and half the mass of the atom.

The series (2.5) could be continued, and one would find that it contains only sums of even powers of the frequencies which can all be expressed similarly to (2.6) in terms of the force constants. However, from the last form of (2.2) it is evident that the series in powers of β will cease to converge when for some frequencies $\hbar\omega > \pi kT$.

In the opposite limit of low temperatures, it would not be right to approximate to (2.4) by assuming $\hbar\omega$ to be large compared with kT, since at no reasonable temperatures is this true for all frequencies. However, one sees that the only frequencies which will give appreciable contributions to (2.4) are those for which $\beta\hbar\omega$ is not large, and, if the temperature is low, this will be true only for the lowest part of the frequency spectrum. Now we have found earlier that the lowest frequencies belong to small values of $\mathbf{f}$, and, if $r > 1$, only to three out of the $3r$ possible frequencies for each $\mathbf{f}$. We also saw that for these vibrations the frequency is proportional to the magnitude of $\mathbf{f}$, the factor of proportionality depending on the direction of $\mathbf{f}$, and on the polarization of the wave.

Hence we may replace the frequency spectrum in (2.4) by

$$\omega(\mathbf{f},s) = c_s(\theta,\phi)f, \qquad s = 1, 2, 3. \qquad (2.7)$$

§ 2.1   CRYSTAL LATTICES. APPLICATIONS

Here θ, ϕ indicate the direction of $\mathbf{f}$, and c_s is a velocity of sound. We may also replace the summation over $\mathbf{f}$ by an integration, noting that the density of permitted values in $\mathbf{f}$ space is $V/(2\pi)^3$. In polar coordinates the integral becomes

$$E_T = \frac{V}{(2\pi)^3} \sum_{s=1}^{3} \int_0^\infty f^3\, df \int\!\!\int d\Omega \frac{\hbar c_s(\theta,\phi)}{e^{\beta \hbar c_s f}-1}, \qquad (2.8)$$

where $d\Omega$ is the element of solid angle. The integration over f may be extended to infinity, since large values do not contribute to the integral. We now introduce a new variable of integration

$$x = \beta \hbar c_s(\theta,\phi) f \qquad (2.9)$$

in place of f.

$$E_T = \frac{V(kT)^4}{(2\pi\hbar)^3} \sum_s \int\!\!\int \frac{d\Omega}{c_s^3} \int_0^\infty \frac{x^3}{e^x-1}\, dx. \qquad (2.10)$$

The last integral can be evaluated and is $\pi^4/15$. The summation over s and integration over angles can be carried out in principle if the elastic constants of the crystal are known. If we put

$$\sum_s \int\!\!\int \frac{d\Omega}{\{c_s(\theta,\phi)\}^3} = \frac{12\pi}{c_{\text{eff}}^3}, \qquad (2.11)$$

then c_{eff} represents a kind of mean value of the sound velocity. With this, finally,

$$E_T = \frac{\pi^2 (kT)^4 V}{10 \hbar^3 c_{\text{eff}}^3}. \qquad (2.12)$$

This T^4 law for the thermal energy, corresponding to a T^3 law for the specific heat at low temperatures, is well confirmed by experiment. It was first derived by Debye, who also gave an interpolation formula which combines the high-temperature law with the low-temperature expression, by using a simplified vibration spectrum. He assumed that the sound velocity is constant for all wavelengths, and independent of direction or polarization, so that the law (2.7) is always true, with constant c_s. Instead of integrating $\mathbf{f}$ over the basic cell of the reciprocal lattice and adding the other types of wave (if $r > 1$), Debye extends the integration over a sphere in $\mathbf{f}$ space, which is chosen in such a way that it gives the right number of degrees of freedom. This means that its radius f_0 is given by

$$\frac{4\pi}{3} f_0^3 \frac{V}{(2\pi)^3} = Nr,$$

or

$$f_0^3 = \frac{6\pi^2 Nr}{V}. \qquad (2.13)$$

The maximum frequency of this model is then

$$\omega_0 = cf_0, \qquad (2.14)$$

c being a constant.

With these assumptions, (2.8) follows at all temperatures, provided the integration over **f** is taken only to the upper limit f_0. Then we obtain in place of (2.10)

$$E_T = \frac{V(kT)^4}{(2\pi\hbar)^3} \frac{12\pi}{c^3} \int_0^{x_0} \frac{x^3}{e^x - 1}\, dx, \qquad (2.15)$$

where $x_0 = \beta\hbar\omega_0 = \hbar\omega_0/kT$. If we introduce a characteristic temperature

$$\Theta = \frac{\hbar\omega_0}{k},$$

the upper limit of the integral in (2.15) is Θ/T. We can also eliminate c by means of (2.13), (2.14):

$$E_T = 9NrkT\left(\frac{T}{\Theta}\right)^3 \int_0^{\Theta/T} \frac{x^3}{e^x - 1}\, dx. \qquad (2.16)$$

This is the Debye formula for the thermal energy of a crystal. From its derivation it is clear that it can only serve as a qualitative result, and as an interpolation formula linking the high- and low-temperature data. The characteristic temperature Θ is treated as an empirical parameter. The law (2.16) is of the right form at low temperatures, where it correctly gives the variation of E_T as T^4, and at very high temperatures, where it gives the Dulong-Petit value of the specific heat, $3k$ per atom. However, it will not in general give the right value of the constant term in (2.5), which is identical with the zero-point energy, nor the coefficient of the T^{-1} term which gives the first deviation of the specific heat from the Dulong-Petit value.

In a modified version of the Debye model, one applies the assumption of constant sound velocity only for the three lowest normal frequencies belonging to each **f**, which form the so-called 'acoustical branch' of the vibration spectrum. The rest form what are sometimes called the 'optical' branches, since for ionic crystals they contain the frequencies which appear in the infra-red absorption spectrum, as we shall see. Since these may be pictured as vibrations of the different parts of the unit cell relative to each other, they are somewhat less dependent on whether adjacent unit cells vibrate with the same or with different phases. Hence these frequencies vary somewhat less strongly with **f** than those in the acoustical branch. For this reason a model is sometimes used in which one replaces the optical branches by $r-1$ frequencies, each belonging to $3N$ different normal modes. For certain cases, e.g. a molecular lattice with weak forces between the molecules as in H_2, this is a more reasonable picture than the simple Debye formula. On the other hand, for an ionic lattice like KCl, in which the mass difference between the two ions is small, it is probably more reasonable to regard the optical branch as a continuation of the acoustic branch than as having constant frequency.

In the older literature one often finds the Debye law in one or the other form taken much more seriously than the derivation justifies, and therefore surprise was caused when precise low-temperature measurements showed considerable deviations from this simple theory. Careful discussion of the vibration spectra of crystals by Blackman and others† showed that the observed deviations were no greater than should be expected.

Nevertheless, the Debye model serves a useful purpose in summarizing the behaviour of E_T qualitatively, and in defining the characteristic temperature which also gives a measure where, for each substance, quantum corrections begin to be important.

2.2. Anharmonic terms. Thermal expansion

We now turn to the discussion of those phenomena for which the cubic terms in (1.20) are no longer negligible. The most important problems of this type concern (a) thermal expansion, (b) specific heat at high temperatures, and (c) thermal conductivity.

For the thermal expansion the most obvious definition of the problem would be in terms of boundary conditions which specify the forces on the surface of the crystal (or the absence of such forces) but allow its dimensions to adjust themselves. In terms of the variables which we have used, this would be very inconvenient, but it is sufficient to calculate the properties of the crystal as a function of temperature and volume, keeping the volume as one of the independent variables.

We may then consider the (Helmholtz) free energy of a crystal of volume $(1+\epsilon)$ times its equilibrium volume at temperature T as a function $F(T, \epsilon)$. For small ϵ we expand:

$$F(T, \epsilon) = F_0(T) + \epsilon F_1(T) + \tfrac{1}{2}\epsilon^2 F_2(T) \qquad (2.17)$$

to second order. Now $\partial F/\partial \epsilon$ is proportional to the pressure and therefore the volume for which the pressure vanishes is given by

$$\frac{\partial F}{\partial \epsilon} = F_1(T) + \epsilon F_2(T) = 0; \qquad (2.18)$$

† Blackman (1935), Kellerman (1940).

hence the actual volume exceeds that for which the potential energy is least by the relative amount

$$\epsilon(T) = -\frac{F_1(T)}{F_2(T)}. \tag{2.19}$$

Hence the numerator F_1 is small at low temperatures, and would, in fact, be zero if the potential were harmonic. On the other hand, the denominator F_2, which, according to (2.17), represents the bulk modulus at temperature T, is finite at $T = 0$ and in the absence of anharmonic forces. It is clear that F_2 will differ only little from its value at low temperatures, and usually we may regard it as constant.

We must now discuss F_1, i.e. consider the behaviour of a crystal if its volume differs from the normal volume, retaining only first-order terms in the expansion.

Since in the series (1.20) for the potential energy the displacements were counted from the sites of the equilibrium lattice, we now introduce new displacements $\mathbf{u}'$ which refer to a similar lattice of a slightly increased lattice constant:

$$\mathbf{u}_{j,\mathbf{n}} = \tfrac{1}{3}\epsilon(\mathbf{d}_j + \mathbf{a}_\mathbf{n}) + \mathbf{u}'_{j,\mathbf{n}}. \tag{2.20}$$

In a general lattice, the uniform expansion will not leave the atoms in an equilibrium position, since for equilibrium the relative positions of different atoms in the unit cell would have to change. We can, however, assert that each atom will still be in equilibrium either if there is only one atom in the unit cell ($r = 1$), or if the relative positions of all atoms in the unit cell are determined by the crystal symmetry, as in the case of the NaCl structure or the hexagonal close-packed lattice. We therefore restrict the discussion to lattices which satisfy this condition. We then know that, if we put all the $\mathbf{u}'$ equal to zero, every atom is in equilibrium, so that the potential energy as a function of the $\mathbf{u}'$ must again be of the form (1.20).

Imagine now (2.20) inserted in (1.20), and the result sorted according to powers of the $\mathbf{u}'$. The terms independent of $\mathbf{u}'$ represent the equilibrium energy of the new lattice. The 'A' terms will contribute to this a term proportional to ϵ^2, which is the elastic energy discussed in (1.57) and contained in the F_2 term of (2.17). Similarly, the 'B' terms contribute a term proportional to ϵ^3, which we may neglect.

The terms linear in the $\mathbf{u}'$ must vanish, since we know the atoms are in equilibrium when $\mathbf{u}' = 0$. We are therefore, to this order, left with an expression of the form (1.20)

$$U - U_0 - \tfrac{1}{2}k\epsilon^2 = \sum_{j,j'} \sum_{\mathbf{n},\mathbf{n}'} \tfrac{1}{2}\mathbf{A}'_{j,\mathbf{n};j',\mathbf{n}'}\, \mathbf{u}'_{j,\mathbf{n}}\, \mathbf{u}'_{j',\mathbf{n}'} +$$
$$+ \sum_{j,j',j''} \sum_{\mathbf{n},\mathbf{n}',\mathbf{n}''} \tfrac{1}{6}\mathbf{B}'_{j,\mathbf{n};j',\mathbf{n}';j'',\mathbf{n}''}\, \mathbf{u}'_{j,\mathbf{n}}\, \mathbf{u}'_{j',\mathbf{n}'}\, \mathbf{u}'_{j'',\mathbf{n}''}, \tag{2.21}$$

§ 2.2 CRYSTAL LATTICES. APPLICATIONS

where the cubic term is still negligible, but the coefficients of the quadratic term are modified:

$$A'_{j,n\,;j',n'} = A_{j,n\,;j',n'} + \tfrac{1}{3}\epsilon \sum_{j'',n''} B_{j,n\,;j',n'\,;j'',n''}(a_{j,n''}+d_{j''}). \qquad (2.22)$$

The occurrence of $a_{n''}$ would seem to introduce a dependence on the choice of that unit cell which we label $n = 0$. Since the uniform expansion does not destroy the translation symmetry of the crystal, this dependence cannot be real; if we change the origin by any lattice vector, the result must remain the same. This requires the condition

$$\sum_{j'',n''} B_{j,n\,;j',n'\,;j''\,n''} = 0, \qquad (2.23)$$

which is similar to (1.22).

We see, then, that the dynamical problem defined by the potential (2.19) is quite similar to the original one, except that, because of the anharmonic terms, all constants defining the vibration spectrum will be changed by corrections of first order in ϵ. The consequences of this are particularly evident if the vibration spectrum can be expressed in terms of a single parameter, Θ, as in the Debye model.

In that case we can write

$$F(T) = -Tg(T/\Theta), \qquad (2.24)$$

where g is a universal function. Indeed, because of the thermodynamic identity

$$E = -T^2 \frac{d}{dT}\left(\frac{F}{T}\right), \qquad (2.25)$$

this leads to

$$E = T\frac{T}{\Theta}g'\left(\frac{T}{\Theta}\right), \qquad (2.26)$$

where g' is the derivative of g with respect to its argument. In other words, E/T is again a universal function of the ratio T/Θ, and this agrees with (2.16). In this case, the dependence on volume is contained in the parameter Θ, and so

$$F_1 = \frac{\partial F}{\partial \epsilon} = T\frac{T}{\Theta}g'\left(\frac{T}{\Theta}\right)\frac{1}{\Theta}\frac{\partial \Theta}{\partial \epsilon}. \qquad (2.27)$$

By comparison with (2.26),

$$F_1(T) = E(T)\frac{1}{\Theta}\frac{\partial \Theta}{\partial \epsilon}. \qquad (2.28)$$

Since we have seen that the temperature dependence of the denominator of (2.19) is negligible, we see that in the Debye model the thermal expansion is proportional to the thermal energy. This law, which was formulated by Grüneisen, is based on an idealized model. Its range of validity is somewhat more general than that of Debye's law, since it holds for a more complicated spectrum provided that, if the body is expanded, all frequencies are changed by the same factor.

We have seen before (cf. (2.10), (2.11), and (2.16)) that at very low temperatures the Debye law becomes exact if we define Θ by

$$(k\Theta)^3 = \frac{3Nr}{V} 2\pi^2 \hbar^3 c_{\text{eff}}^3, \qquad (2.29)$$

where c_{eff} is defined in (2.11). In that case (2.28) becomes

$$F_1(T) = E(T)\left(-\frac{1}{3} + \frac{\partial}{\partial \epsilon} \log c_{\text{eff}}\right). \qquad (2.30)$$

It therefore follows that the thermal expansion at low temperatures is proportional to T^4. It is also easy to see that at high temperatures ($T \gg \Theta$) the expansion becomes linear in T, i.e. the expansion coefficient becomes constant as the Grüneisen formula (2.28) predicts. This, of course, holds only apart from higher-order corrections, such as cubic and higher-order terms in the expansion (2.17), and the variation of the denominator of (2.19).

Lastly, it should be mentioned that the method used here will, in non-cubic crystals, still give the variation of free energy with volume, but will not give the thermal expansion correctly, since in general the shape of the crystal also changes with temperature. In such cases (2.17) has to be replaced by a function of several strain components, and (2.18) is replaced by several simultaneous equations. The Grüneisen law remains valid for the expansion in each direction if the vibration spectrum retains its shape under an arbitrary shear strain.

2.3. Linear term in specific heat

A second phenomenon entirely dependent on the anharmonic nature of the forces is the deviation of the specific heat from the Dulong-Petit value at high temperatures. The anharmonic terms will also cause a small correction to the specific heat at lower temperatures, but, since we have seen that the theoretical law cannot in practice be found to high accuracy, small corrections are of no practical interest there. On the other hand, at high temperatures where quantum effects are negligible, the theory with harmonic forces gives a simple answer, and small deviations from this can be compared with experiment.

For this purpose it is evidently sufficient to give only a classical treatment. Before writing down the general equations, it may help to discuss

§ 2.3 CRYSTAL LATTICES. APPLICATIONS

the corresponding problem for a single particle oscillating in a potential field. Here the Hamiltonian form of the energy is

$$H = \frac{1}{2m}p^2 + V(x), \qquad (2.31)$$

and the partition function is therefore

$$Z(\beta) = \int dp \int dx \, e^{-\beta p^2/2m - \beta V(x)}. \qquad (2.32)$$

As usual, the integration over p can be carried out immediately, leaving

$$Z(\beta) = \left(\frac{2\pi m}{\beta}\right)^{\frac{1}{2}} \int dx \, e^{-\beta V(x)}. \qquad (2.33)$$

Now let the potential be approximately that of a harmonic oscillator, with small anharmonic terms,

$$V(x) = \tfrac{1}{2}ax^2 + bx^3 + cx^4. \qquad (2.34)$$

Then, if we expand the integral in (2.33) in powers of b and c, retaining the first-order term in c and the second-order term in b,

$$Z(\beta) = \left(\frac{2\pi m}{\beta}\right)^{\frac{1}{2}} \int dx \, e^{-\frac{1}{2}\beta a x^2}(1 - \beta b x^3 + \tfrac{1}{2}\beta^2 b^2 x^6 - \beta c x^4). \qquad (2.35)$$

Here the integral over the cubic term evidently vanishes by symmetry. The remaining integrations are elementary, and the result is

$$Z(\beta) = 2\pi \sqrt{\left(\frac{m}{a}\right)} \frac{1}{\beta} \left(1 - \frac{3c}{a^2 \beta} + \frac{15 b^2}{2a^3 \beta}\right). \qquad (2.36)$$

Now we obtain the mean thermal energy from the partition function by using the relation

$$E = -\frac{\partial \log Z}{\partial \beta} \qquad (2.37)$$

which gives, neglecting again higher-order terms,

$$E = kT - \frac{3c}{a^2}(kT)^2 + \frac{15 b^2}{2a^3}(kT)^2. \qquad (2.38)$$

To retain the analogy with the forces in a crystal we assume that each derivative of the potential differs from the next higher one by a factor of the order of the atomic distance, and this means that in order of magnitude a/b is comparable to b/c. In that case the last two terms of (2.38) are of comparable magnitude. Both give corrections to the energy proportional to T^2, hence linear terms in the specific heat. If in order of magnitude $a/b \sim b/c \sim d$, then both the second and the third term are of order $(kT)^2/ad^2$. In other words, they are small compared with the first, as long as the temperature is small compared with that temperature for which the mean square amplitude is d^2. One easily verifies that the terms of higher order which we have neglected, i.e. higher-order terms in the potential (2.34), higher terms in the expansion (2.35) of the exponential and in the expansion of the logarithm in (2.37), contain higher powers of kT/ad^2, so that (2.38) is a consistent approximation.

After this preparation we turn to the general case. It is clear that we must again include the fourth-order terms not explicitly shown in (1.20). Just as in (2.35) we shall again expand the Boltzmann factor in powers of the cubic and quartic terms, retaining the part proportional to the square of the cubic and that proportional to the quartic terms.

Since, however, the exponential contains the quadratic (harmonic) terms, we must use variables in terms of which the quadratic terms separate. In other words, we must use normal coordinates. We therefore substitute in the **B** term of (1.20) the expression (1.30) for the u:

$$U_c = \tfrac{1}{6}\sum_{\mathbf{f},\mathbf{f}',\mathbf{f}''}\sum_{s,s',s''} b(\mathbf{f},s;\mathbf{f}',s';\mathbf{f}'',s'') q_{\mathbf{f},s} q_{\mathbf{f}',s'} q_{\mathbf{f}'',s''}, \qquad (2.39)$$

where

$$b(\mathbf{f},s;\mathbf{f}',s';\mathbf{f}'',s'')$$
$$= \sum_{j,j',j''}\sum_{\mathbf{n},\mathbf{n}',\mathbf{n}''} \mathbf{B}_{j,\mathbf{n};j',\mathbf{n}';j'',\mathbf{n}''}\, e^{i(\mathbf{f}\cdot\mathbf{a}_\mathbf{n}+\mathbf{f}'\cdot\mathbf{a}_{\mathbf{n}'}+\mathbf{f}''\cdot\mathbf{a}_{\mathbf{n}''})} \mathbf{v}_j(\mathbf{f},s)\mathbf{v}_{j'}(\mathbf{f}',s')\mathbf{v}_{j''}(\mathbf{f}'',s''). \qquad (2.40)$$

From the translational symmetry follows at once an important property of the b coefficients. We know that the equations remain unchanged if we displace the crystal by a lattice vector, which means adding a lattice vector $\mathbf{a}$ to all the $\mathbf{a}_\mathbf{n}$ in (2.40). This multiplies the expression by

$$e^{i\mathbf{a}\cdot(\mathbf{f}+\mathbf{f}'+\mathbf{f}'')}. \qquad (2.41)$$

Since the expression must remain unchanged, it follows that b must vanish, unless the exponential (2.41) is unity, which by (1.28) means that

$$b(\mathbf{f},s;\mathbf{f}',s';\mathbf{f}'',s'') = 0, \qquad (2.42)$$

unless $\mathbf{f}+\mathbf{f}'+\mathbf{f}'' = \mathbf{K}$ is a vector of the reciprocal lattice. This means that, if two of the vectors, say $\mathbf{f}, \mathbf{f}'$, are given, all values of $\mathbf{f}''$ for which the coefficient is non-zero are equivalent, and only one of them lies in the basic cell of the reciprocal lattice, which by definition contains all our wave vectors $\mathbf{f}$. If $\mathbf{f}$ and $\mathbf{f}'$ are small, so that their sum lies itself in the basic cell, we must take $\mathbf{K} = 0$ in (2.42), otherwise it must be chosen so as to bring $\mathbf{f}''$ back into the basic cell.

§ 2.3 CRYSTAL LATTICES. APPLICATIONS

We now write for the Boltzmann factor†

$$e^{-\beta E} = e^{-\beta E_{\text{kin}} - \beta U_h}(1 - \beta U_c + \tfrac{1}{2}\beta^2 U_c^2 - \beta U_q) \tag{2.43}$$

as in (2.35), where U_h, U_c, U_q stand for the harmonic, cubic, and quartic parts of the potential energy respectively. To obtain the partition function Z we have to integrate (2.43) over phase space, which amounts to taking the average values of U_c, U_c^2, and U_q over the thermal distribution appropriate to harmonic motion. Now U_h is even in all the normal coordinates; hence the probability is not altered by reversing the value of all the q, and therefore the average of U_c, which is an odd function of the q, vanishes.

For the discussion of the term in U_c^2 we note that the normal coordinates are statistically independent and that the average potential energy for each of them must be $\tfrac{1}{2}kT = 1/2\beta$. Hence by (1.47)

$$\overline{|q_{\mathbf{f},s}|^2} = \{M^{(N)}\beta\omega^2(\mathbf{f},s)\}^{-1}, \tag{2.44}$$

where the bar over the symbol means the average value for harmonic motion.

Now the square of (2.39) contains products of the type (we form the product $U_c^* U_c$, which is the same since U_c is real)

$$q_{\mathbf{f},s}^* q_{\mathbf{f}',s'}^* q_{\mathbf{f}'',s''}^* q_{\mathbf{f}_1,s_1} q_{\mathbf{f}_2,s_2} q_{\mathbf{f}_3,s_3}. \tag{2.45}$$

Because of the statistical independence of the q, the average of this product vanishes, unless the six normal modes concerned are equal in pairs. The symmetry of the coefficients in (2.39) means that we can permute the first three between themselves and the last three between themselves without changing the expression. Hence there are only two types of term to be considered:

$$\left. \begin{array}{l} (a) \quad \mathbf{f} = -\mathbf{f}', \mathbf{f}_1 = -\mathbf{f}_2, \mathbf{f}'' = \mathbf{f}_3; \; s = s', \; s_1 = s_2, \; s'' = s_3 \\ (b) \quad \mathbf{f} = \mathbf{f}_1, \mathbf{f}' = \mathbf{f}_2, \mathbf{f}'' = \mathbf{f}_3; \; s = s_1, \; s' = s_2, \; s'' = s_3 \end{array} \right\}. \tag{2.46}$$

At this point the further discussion is greatly simplified if we are dealing with a lattice for which the unit cell contains only one atom ($r = 1$), and I shall restrict myself to this case.‡ Now, in the case (a), (2.42) requires that $\mathbf{f}'' = \mathbf{f}_3 = 0$. If this is inserted in (2.40) for $r = 1$,

---

† The expansion of the exponential which has been used in deriving (2.43) looks doubtful, since for the whole crystal neither U_c nor U_q would be expected to be small compared with kT. However, a more rigorous justification of (2.43) can be given, based on the fact that in any configuration which occurs with reasonable probability that part of U_c or U_q which depends on any given normal coordinate is small, so that, in averaging over each coordinate, the expansion may be used.

‡ In a general lattice terms of the type (a) can be shown to be connected with the variation of the cell structure with temperature. In particular, they still vanish if the positions d_j of the atoms in a cell are determined by the crystal symmetry.

when the suffix j can be omitted, we see that the appropriate coefficient vanishes by (2.23). It represents the change in the harmonic terms caused by a uniform displacement of the lattice.

We are therefore left with case (b) and the equivalent ones obtained by permuting the first three normal modes. Hence

$$\tfrac{1}{2}\beta^2 U_c^2 = \frac{1}{12}\frac{1}{M^{(N)3}\beta}\sum_{\mathbf{f},\mathbf{f}',\mathbf{f}''}\sum_{s,s',s''}\frac{|b(\mathbf{f},s;\mathbf{f}',s';\mathbf{f}'',s'')|^2}{\omega^2(\mathbf{f},s)\omega^2(\mathbf{f}',s')\omega^2(\mathbf{f}'',s'')}. \quad (2.47)$$

Strictly speaking, this expression should be corrected for the case when two of the three normal modes in the sum coincide, since we are then concerned with the average of the fourth power of one of the normal coordinates, which is not the same as the square of the mean square. However, the contribution of such terms is of relative magnitude $1/N$, and hence negligible. The contribution to the energy can again be obtained from (2.37) and, neglecting higher-order corrections, yields

$$E_c = \frac{(kT)^2}{12 M^{(N)3}}\sum_{\mathbf{f},\mathbf{f}',\mathbf{f}''}\sum_{s,s',s''}\frac{|b(\mathbf{f},s;\mathbf{f}',s';\mathbf{f}'',s'')|^2}{\omega^2(\mathbf{f},s)\omega^2(\mathbf{f}',s')\omega^2(\mathbf{f}'',s'')}. \quad (2.48)$$

Because of the restriction (2.42), there are $3^3 N^2$ terms in this sum. When the sum in (2.40) does not vanish, it is proportional to N. Hence, allowing for the mass denominator, (2.48) is proportional to N, as it should be. It is proportional to T^2, as the simple model led us to expect. One might suspect that the frequency denominators would lead to trouble, since the lowest frequencies are practically zero. However, they belong to the acoustical limit when $\mathbf{f}$ is very small, and in that case the numerator is also small, since we had already seen that then $b(\mathbf{f},s;\mathbf{f}',s';\mathbf{f}'',s'')$ vanishes if $\mathbf{f}=0$; it will also be small if $\mathbf{f}$ is small. The same applies for small $\mathbf{f}'$ and $\mathbf{f}''$.

For a better interpretation of the sum (2.48) it is therefore convenient to define a new coefficient $\mathfrak{b}$ by

$$b(\mathbf{f},s;\mathbf{f}',s';\mathbf{f}'',s'')$$
$$=\omega(\mathbf{f},s)\omega(\mathbf{f}',s')\omega(\mathbf{f}'',s'')M^{(N)}\Delta(\mathbf{f}+\mathbf{f}'+\mathbf{f}'')\mathfrak{b}(\mathbf{f},s;\mathbf{f}',s';s''). \quad (2.49)$$

Here $\Delta(\mathbf{f})$ is a generalized Kronecker symbol for which

$$\Delta(\mathbf{f}) = \begin{cases} 1, & \text{if } \mathbf{f} \text{ is a vector in the reciprocal lattice,} \\ 0, & \text{otherwise.} \end{cases} \quad (2.50)$$

The factor $M^{(N)}$ is inserted to make $\mathfrak{b}$ independent of the size of the crystal. $\mathfrak{b}$ then approaches a finite limit when any of the frequencies vanishes, and may be supposed to have constant order of magnitude for all arguments.

Then the cubic contribution to the thermal energy becomes finally

$$E_c = \frac{(kT)^2}{12M^{(N)}} \sum_{\mathbf{f},\mathbf{f}'} \sum_{s,s',s''} |b(\mathbf{f},s;\mathbf{f}',s';s'')|^2. \tag{2.51}$$

For the contribution of the quartic terms, we put these in a form analogous to (2.49)

$$U_q = \tfrac{1}{24} \sum c(\mathbf{f},s;\mathbf{f}',s';\mathbf{f}'',s'';\mathbf{f}''',s''') q_{\mathbf{f},s}\, q_{\mathbf{f}',s'}\, q_{\mathbf{f}'',s''}\, q_{\mathbf{f}''',s'''}. \tag{2.52}$$

Again only those terms contribute to the average for which the normal modes are equal in pairs. They can be grouped in pairs in three ways and we therefore have

$$\beta \overline{U}_q = \frac{1}{8} \sum_{\mathbf{f},\mathbf{f}'} \sum_{s,s'} \frac{c(\mathbf{f},s;\mathbf{f},s;\mathbf{f}',s';\mathbf{f}',s')}{\{M^{(N)}\}^2 \beta \omega^2(\mathbf{f},s)\omega^2(\mathbf{f}',s')}. \tag{2.53}$$

We conclude, as before, that $c(\mathbf{f},s;...)$ must vanish if any of its wave vectors tends to zero. Also, its non-vanishing elements will again contain a factor proportional to the size of the crystal. Hence, if we define

$$c(\mathbf{f},s;\mathbf{f}',s';\mathbf{f}'',s'';\mathbf{f}''',s''')$$
$$= M^{(N)}\omega(\mathbf{f},s)\omega(\mathbf{f}',s')\omega(\mathbf{f}'',s'')\omega(\mathbf{f}''',s''') \times$$
$$\times \Delta(\mathbf{f}+\mathbf{f}'+\mathbf{f}''+\mathbf{f}''') \mathfrak{c}(\mathbf{f},s;\mathbf{f}',s';\mathbf{f}'',s'';s'''), \tag{2.54}$$

the coefficient $\mathfrak{c}$ will be independent of the crystal size, and roughly constant in order of magnitude.

The contribution of the quartic terms to the thermal energy is now

$$E_q = -\frac{(kT)^2}{8M^{(N)}} \sum_{\mathbf{f},\mathbf{f}'} \sum_{s,s'} \mathfrak{c}(\mathbf{f},s;-\mathbf{f},s;-\mathbf{f}',s';s'), \tag{2.55}$$

which again has the same temperature dependence as (2.51). One can also show that the two contributions are of the same order of magnitude.

These expressions could easily be generalized to their quantum-mechanical form, but this is not worth while since they are of no interest at low temperatures.

The results (2.51) and (2.55) represent the correction to the Dulong-Petit value of the specific heat if the volume of the crystal is kept constant at the equilibrium value. In actual fact, the crystal will expand as the temperature increases. This will result in a further correction term, which can be calculated from the thermal expansion coefficient and the elastic constants by using purely thermodynamic arguments. I shall not give this calculation here.

2.4. Thermal conductivity

We have seen that the thermal energy of the crystal consists of waves which travel with velocities of the order of sound velocity. They are therefore also capable of transporting energy. Using the familiar argument of the kinetic theory of gases, we may write the dimensional formula
$$\kappa = \gamma C c \lambda, \qquad (2.56)$$
where κ is the heat conductivity, C the specific heat per unit volume, c the velocity of sound, λ a mean free path, and γ a numerical factor.

In the harmonic approximation the waves travel freely without attenuation, and therefore have an unlimited free path. In reality their free path is limited by (a) the anharmonic terms, (b) impurities and imperfections in the crystal, and (c) the finite dimensions of the crystal. We shall for the present consider only the first effect, which is dominant in good and pure crystals of not too small size at not too low temperatures.

For this problem we shall use quantum mechanics at once, since the case of low temperatures is of interest. The classical result will, of course, also come out as a limit. We must now regard the cubic part of the energy as a perturbation, and we may take it in the form (2.39). Since we found the most convenient variables for the quantum treatment to be the Q introduced by (1.32), we write

$$U_c = \tfrac{1}{6} \sum_{\mathbf{f},\mathbf{f}',\mathbf{f}''} \sum_{\sigma,\sigma',\sigma''} b(\mathbf{f},\sigma;\mathbf{f}',\sigma';\mathbf{f}'',\sigma'') Q(\mathbf{f},\sigma) Q(\mathbf{f}',\sigma') Q(\mathbf{f}'',\sigma''), \qquad (2.57)$$

where the sums over σ now cover positive and negative values.

Now we consider (2.57) as a small perturbation which causes transitions between the states of the unperturbed system. The general relation giving the transition probability per unit time from some initial state i to a final state f is[†]

$$\frac{dP_{if}}{dt} = \frac{2\pi}{\hbar} |(i|U_c|f)|^2 \delta(E_f - E_i). \qquad (2.58)$$

Here δ is the Dirac delta-function, E_i, E_f are the energy values of initial and final state respectively, and $(i|U_c|f)$ is the matrix element of (2.57) between the two states. Now we have seen in (1.63) that, for positive σ, Q has matrix elements only corresponding to an increase of the corresponding quantum number by 1, for negative σ to a decrease. Each term in (2.57) therefore describes a process in which three quantum numbers each change by one unit. The corresponding change of energy is

$$\hbar\{\omega(\mathbf{f},\sigma) + \omega(\mathbf{f}',\sigma') + \omega(\mathbf{f}'',\sigma'')\}, \qquad (2.59)$$

[†] See, for example, Schiff (1949), section 29.

§ 2.4   CRYSTAL LATTICES. APPLICATIONS

and for the process to be possible according to (2.58) this must vanish. Since the sign of $\omega(\mathbf{f}, \sigma)$ is the sign of σ, it is clear that σ, σ', σ'' cannot be all positive or all negative. The actual process therefore involves either two positive and one negative, or vice versa, which means either the destruction of two phonons and the creation of a new one, or vice versa. Hence the process in which two phonons $\mathbf{f}, s$ and $\mathbf{f}', s'$ are combined into a single $\mathbf{f}'', s''$ has the probability (allowing for permutations in (2.57))

$$|b(\mathbf{f}, s; \mathbf{f}', s'; -\mathbf{f}'', s'')|^2 \frac{2\pi\hbar N(\mathbf{f}, s) N(\mathbf{f}', s')\{N(\mathbf{f}'', s'')+1\}}{8\{M^{(N)}\}^3 \omega(\mathbf{f}, s)\omega(\mathbf{f}', s')\omega(\mathbf{f}'', s'')} \times$$

$$\times \delta\{\omega(\mathbf{f}, s) + \omega(\mathbf{f}', s') - \omega(\mathbf{f}'', s'')\}. \quad (2.60)$$

The δ-function vanishes unless

$$\omega(\mathbf{f}, s) + \omega(\mathbf{f}', s') - \omega(\mathbf{f}'', s'') = 0, \quad (2.61)$$

and by (2.42) the first factor vanishes unless also

$$\mathbf{f} + \mathbf{f}' - \mathbf{f}'' = \mathbf{K}, \quad (2.62)$$

where $\mathbf{K}$ again stands for a lattice vector of the reciprocal lattice.

The first of these conditions expresses the conservation of energy in the collision between phonons. The second is similar to conservation of momentum. Indeed, if we were dealing with a continuous medium instead of a crystal, the only possible value of $\mathbf{K}$ on the right-hand side of (2.62) would be zero, and in a continuous medium the momentum of a phonon of wave vector $\mathbf{f}$ is $\hbar\mathbf{f}$. However, this analogy must not be taken too seriously, since for the normal vibrations which we have derived the total momentum certainly vanishes (except for the rather degenerate case $\mathbf{f} = 0$).

With our convention, which restricts the $\mathbf{f}$ values to the basic cell of the reciprocal lattice, there exists for given $\mathbf{f}$ and $\mathbf{f}'$ always just one value of $\mathbf{f}''$ which satisfies (2.62); whether the $\mathbf{K}$ on the right is a finite reciprocal lattice vector or zero will depend on the choice of $\mathbf{f}$ and $\mathbf{f}'$.

Processes for which $\mathbf{K} \neq 0$ in (2.62) are called 'Umklapprozesse' in the German literature. A typical example is the case when $\mathbf{f}$ is a wave travelling in the $+x$ direction, almost at the edge of the basic cell, i.e. with a wavelength such that adjacent atoms move with almost the same phase in opposite directions. The wave $\mathbf{f}'$ which interacts with it has a very small wave vector, also in the $+x$-direction, but enough to bring the sum beyond the edge of the basic cell, so that it is equivalent to a vector near the left-hand face of the cell, $\mathbf{K}$ being in that case the smallest non-zero reciprocal lattice vector in the x-direction.

It will be clear from this example that physically there is no important difference between processes in which the sum $\mathbf{f}+\mathbf{f}'$ just remains within the basic cell, and those in which it falls just outside and has to be brought back by adding a suitable $\mathbf{K}$, and indeed the distinction between the two depends on our convention in choosing the basic cell. What matters, however, is whether or not we can find a convention with which (2.62) would always hold with $\mathbf{K} = 0$.

Indeed, let us assume for the moment that the 'Umklapp' processes could be excluded and that we were allowed to put $\mathbf{K} = 0$ in (2.62). Then it is obvious that the sum of the wave vectors would not change in the process we have specified, and that therefore the quantity

$$\mathbf{J} = \sum_{\mathbf{f}} \sum_{s} \mathbf{f} N(\mathbf{f}, s) \qquad (2.63)$$

would be conserved in these collisions. Hence the collisions would be incapable of establishing complete statistical equilibrium if once $\mathbf{J} \neq 0$. Rather they will make the system approach the most probable state which is compatible with a given $\mathbf{J}$. (The statistical problem is similar to that of collisions between the molecules of a gas in an infinitely long straight frictionless tube if the average velocity is non-zero.) Now a resultant $\mathbf{J}$ in the positive x-direction means that the phonons in the positive x-direction will be more numerous than those travelling in the opposite direction, and this asymmetry will lead to a non-vanishing energy transport. In other words, in this hypothetical case a finite energy transport can persist without a temperature gradient to maintain it, and this implies an infinite thermal conductivity.

It follows, therefore, that the solutions of (2.61), (2.62) with $\mathbf{K} \neq 0$ are of vital importance for the thermal resistance.

The consequences of this conclusion are particularly obvious at very low temperatures ($T \ll \Theta$). In that case most of the phonons which make up the thermal motion of the crystal are long-wave ones, and for the interaction of long-wave phonons $\mathbf{K}$ will vanish in (2.62). If the smallest non-zero value of $\mathbf{K}$ is of magnitude K_0, then $\mathbf{f}+\mathbf{f}'$ must in magnitude exceed $\frac{1}{2}K_0$, if the addition of a lattice vector is to lead to a reduction. Hence either $\mathbf{f}$ or $\mathbf{f}'$ must exceed $\frac{1}{4}K_0$ in magnitude. We can thus give a lower limit to the frequency of the phonons that must be present to make an Umklapp process possible. At low temperatures the number $N(\mathbf{f}, s)$ of such phonons is small, and hence the magnitude of the expression (2.60) is governed by the factors $N(\mathbf{f}, s)$, $N(\mathbf{f}', s')$ which will diminish exponentially with temperature if one of the frequencies is kept above a fixed limit. We see therefore that the rate at which

Umklapp processes take place diminishes with temperature as
$$e^{-\gamma\theta/T}, \qquad (2.64)$$
where γ is some numerical factor less than 1. Since in the absence of such processes there is no thermal resistance, the thermal resistance itself must be proportional to (2.64). This conclusion is confirmed by experiments at Oxford.†

It is evident that this result is not altered if we consider also the reverse process, in which one phonon splits in two, since this will produce a non-vanishing **K** in (2.62) only if the frequency of the initial phonon, in this case $\omega(\mathbf{f}'', s'')$, is high enough.

It is equally clear that the situation is not substantially changed if we include the quartic terms in the potential. These will give rise to four-phonon processes. Either two phonons collide and change into two new ones, or three merge into one, or lastly one splits into three. In each case we find conservation laws as above, and we find that the total wave vector (2.63) is conserved except for processes which require the initial presence of at least one high-energy phonon. However, if one were to attempt a closer estimate of the factor in (2.64), one would have to decide whether for this purpose the cubic or the quartic terms are dominant.

For a further understanding of the interactions between phonons it is obviously important to consider the conservation laws (2.61) and (2.62) to see the nature of the solutions which they admit. We shall carry out this discussion for a lattice with one atom per unit cell, for which there are then just three frequencies for any given **f**. Consider, for further simplification, the case in which **f**, **f**′, **f**″ are all parallel to a main axis of the crystal. Then two of the normal modes will in each case be transverse and belong to the same frequency, the third will be longitudinal and have a higher frequency. The frequencies will be represented by curves of the shape of the full lines in Fig. 1. Then we may imagine the solutions of (2.61) and (2.62) constructed graphically as follows: Mark on the full line the point **f**, belonging to one of the initial phonons, and then draw the frequency curves again, with that point as origin (broken lines), continuing them periodically as necessary; then their intersections with the original full lines give the required solution. The abscissa and ordinate of the intersection give **f**″ and ω'', whereas the horizontal and vertical distances of the intersection from the origin of the broken curves give **f**′ and ω''. If by this construction **f**′ falls outside the basic cell, it has to be reduced in the usual way, and we are then dealing with an Umklapp process. These quantities have been labelled in Fig. 1

† Cf. Berman (1951), (1953).

for one particular intersection, in which two transverse phonons combine to make up a longitudinal one. The figure also shows inter-

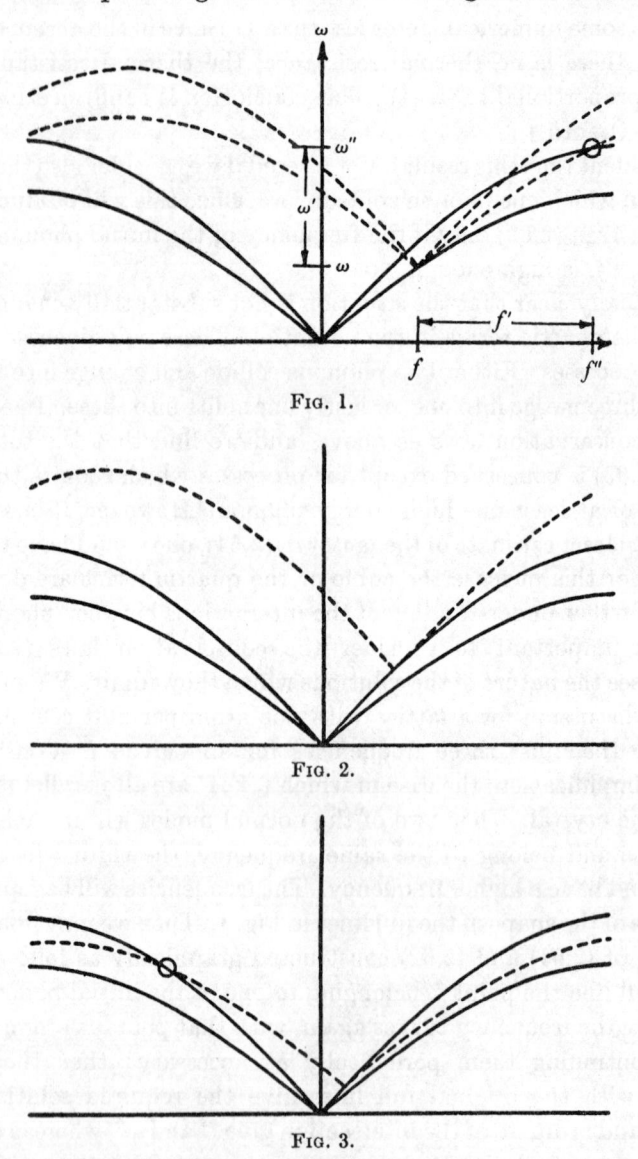

FIG. 1.

FIG. 2.

FIG. 3.

sections corresponding to one transverse and one longitudinal phonon combining into a longitudinal one. The same construction for two initial longitudinal quanta is shown in Fig. 2, and no solutions are obtained in this case. Fig. 3 shows that, if **f** is in particular very small (long waves) and s longitudinal, then the only possible process is a collision with a transverse phonon of comparable wavelength.

All these arguments hold only for the very special direction of the three wave vectors. By imagining the figures extended first into two and then into three dimensions, it is, however, easy to see that, if we denote by $s = 1, 2, 3$ frequencies for given $\mathbf{f}$ in increasing order of magnitude, the following statements are still correct: no solution exists for which all phonons belong to the same s, or for the case

$$(3)+(3) \to \begin{cases} (1), \\ (2); \end{cases}$$

possible processes are

$$(1)+(1) \to (2), \qquad (1)+(1) \to (3), \qquad (1)+(2) \to (2),$$
$$(1)+(2) \to (3), \qquad (1)+(3) \to (3), \qquad (2)+(3) \to (3).$$

In each of these cases we may give one of the initial phonons arbitrarily, and then find a surface on which the second wave vector must end. It is again true that, if in the last two cases we make the wavelength for $s = 3$ very long, the colliding phonon must have a wavelength of the same order.

2.5. Boltzmann equation

We shall now derive the integral equation whose solution would give a quantitative expression for the thermal conductivity, even though we shall not succeed in solving it. This seems worth while because it is possible to draw some qualitative conclusions, and also because it is of some interest to see the nature of the difficulties which stand in the way of a complete solution.

In order to describe heat conduction we must be able to describe a state of affairs in which the temperature is different at different points in space. For this purpose we should not use exact normal coordinates as we have done so far, but rather combinations of them in the form of wave packets, such that each belongs to a small spread $\delta \mathbf{f}$ of the wave vector, and is localized in space in a region of size δx. We have from the uncertainty principle

$$\delta f_x \delta x \sim 1, \qquad (2.65)$$

and similarly for the other components. This means, of course, that we are considering vibrations which are not strictly normal modes, or quantum states which are not strictly stationary. The first effect of this is that each of these wave packets is going to travel with the appropriate group velocity, and this we shall, of course, take into account. Furthermore, the energy of such a wave is not sharply defined, so that we must admit a slight spread in equation (2.61) if the frequencies in that equation are taken to be the mean values for each wave packet. However, this will make no difference in the results if all other energy-dependent factors in the equations are practically constant over the region of this uncertainty. Now in thermal equilibrium, or in the slight deviations from equilibrium that correspond to a reasonable heat flow, the excitation numbers of various phonon states change appreciably only over energy changes of the order of kT. Hence the use of wave packets causes no complication provided

$$kT > \delta E = \hbar \frac{\partial \omega}{\partial f} \delta f = \hbar v\, \delta f,$$

where v is the group velocity. Inserting for δf from (2.65), the condition becomes

$$\delta l > \frac{\hbar v}{kT} \sim \frac{\Theta}{T} a, \qquad (2.66)$$

where in the last estimate we have replaced the group velocity by the sound velocity, and identified the length $1/f_0$ which enters in the definition of the Debye temperature with the lattice constant a. It is therefore sufficient to assume that the temperature should not vary appreciably over distances of the order (2.66), which is evidently a very mild restriction on the temperature gradient. In any case the coefficient of thermal conductivity refers by definition to an infinitesimal gradient; but (2.66) shows that there is no reason to expect in practice any dependence on the magnitude of the gradient from this cause.

We then regard the number of phonons as a function of their wave vector and polarization as well as of the space coordinates, and we formulate the condition for this distribution function to remain stationary. It is subject to change for two reasons: one is the motion of the phonons, carrying energy from the high-temperature to the low-temperature regions, the other is the effect of the collisions between phonons which we have just discussed. The probability of such collisions is given by (2.60), and to work out the average effect of this we should really require to know the average value of products of three occupation numbers, which may be replaced by the product of the averages only in the absence of correlations. In fact, if we assume that the numbers of phonons in different states are initially uncorrelated and then consider their rate of change by (2.60), we find that at a later instant there will be correlations between the numbers for those states between which collisions may have occurred. This is analogous to the situation in the kinetic

§ 2.5   CRYSTAL LATTICES. APPLICATIONS   47

theory of gases where one can show that after a collision the distribution in phase space of the molecules which have just collided will show correlations. This point is then dealt with by remarking that it is most unlikely that the same pair of molecules will collide again before each of them has made many collisions with others, and that it is therefore reasonable to suppose the properties of any two molecules which do collide to be uncorrelated.

The same argument can be taken over in our case, provided conditions are such that we are dealing with large numbers of phonons, so that the chance of the same individual phonons colliding several times in succession is negligible.

There is one further condition to be satisfied before we can write an equation for the mean occupation numbers. This is that the occupation numbers should contain all the information that is obtainable about the system, and that there should be no phase relations between states with different phonon numbers. In terms of the concept of a statistical density matrix this means that the density matrix should be diagonal in the phonon numbers. I shall not here give a formal argument that this is a consistent assumption, i.e. that, inserting a diagonal density matrix for the initial state, we shall find it diagonal at later times, using only the assumption that we are dealing with large numbers of phonons, which we have used already. I want to point out, however, that this statement can be true only if we have chosen variables which are appropriate to the problem. For example, we might have chosen standing waves (sine and cosine waves) instead of the progressive waves which occurred in our definition of the normal modes. In that case a state in which there is an energy transport can be specified only by phase relations between the sine and cosine waves. If we simply omitted all phase relations, and hence all non-diagonal elements in all equations, we should come to the conclusion that a temperature gradient has no tendency to produce an energy flow, which would be patently incorrect. At the same time we should also be unable to recognize the importance of the total wave vector (2.63) whose average value for standing waves is, of course, zero, so that it would again depend on phase relations between different standing waves.

After these remarks we consider the number $N(\mathbf{f}, s, \mathbf{r})$ of phonons in the wave packet of wave vector $\mathbf{f}$, polarization s, and located at $\mathbf{r}$. Then its rate of change due to the motion of the phonons is

$$-\frac{\partial N}{\partial x} v_x$$

if the gradient is in the x-direction. Here v_x is the x-component of the group velocity, which equals $\partial\omega(\mathbf{f},s)/\partial f_x$.

In computing the change due to collisions, we note that (2.60) is independent of the size of the crystal, so that we may retain its exact form even though we are now thinking of localized wave packets, which is equivalent to considering a smaller crystal. Indeed, the rate of change of $N(\mathbf{f},s)$ by the process specified in (2.60) may be written, using (2.49),

$$-\frac{\pi\hbar}{4M^{(N)}}\sum_{\mathbf{f}'}\sum_{s',s''}|\mathfrak{b}(\mathbf{f},s;\mathbf{f}',s';s'')|^2\omega\omega'\omega''\delta(\omega+\omega'-\omega'')NN'(N''+1). \tag{2.67}$$

Here we have written for brevity ω, ω', ω'' in place of $\omega(\mathbf{f},s)$, $\omega(\mathbf{f}',s')$, $\omega(\mathbf{f}'',s'')$ respectively, and N, N', N'' for $N(\mathbf{f},s)$, $N(\mathbf{f}',s')$, $N(\mathbf{f}'',s'')$. We have also assumed that $\mathbf{f}''$ is the solution of (2.62) for given $\mathbf{f}$ and $\mathbf{f}'$. The summation over $\mathbf{f}'$ may now be replaced by an integration. We have seen earlier (§ 1.6) that there are $V/(2\pi)^3$ permissible values of $\mathbf{f}$ per unit volume in f space, if V is the volume of the crystal. Since this is related to the total mass by $M^{(N)} = \rho V$, where ρ is the density, we have finally for (2.67)

$$-\frac{\hbar}{32\pi^2\rho}\sum_{s',s''}\int d^3\mathbf{f}'|\mathfrak{b}(\mathbf{f},s;\mathbf{f}',s';s'')|^2\omega\omega'\omega''\delta(\omega+\omega'-\omega'')NN'(N''+1), \tag{2.68}$$

which is now indeed independent of the size. In addition to this process we have to take into account the inverse (a phonon with wave vector $\mathbf{f}''$ splits into those with $\mathbf{f}$ and $\mathbf{f}'$), and also the process in which the phonon under consideration, $\mathbf{f}$, splits into two others or vice versa. Combining these four processes and the phonon motion, we have finally for the total rate of change of $N(\mathbf{f},s)$

$$-\frac{\partial N}{\partial x}\frac{\partial\omega}{\partial f_x}+\frac{\hbar}{32\pi^2\rho}\int d^3\mathbf{f}'\Big[\sum_{s's''}|\mathfrak{b}(\mathbf{f},s;\mathbf{f}',s';s'')|^2\times$$
$$\times\omega\omega'\omega''\delta(\omega+\omega'-\omega'')\{(N+1)(N'+1)N''-NN'(N''+1)\}+$$
$$+\sum_{s',s''}\tfrac{1}{2}|\mathfrak{b}(\mathbf{f},s;\mathbf{f}',s';s''')|^2\omega\omega'\omega'''\delta(\omega-\omega'-\omega''')\times$$
$$\times\{(N+1)N'N'''-N(N'+1)(N'''+1)\}\Big]. \tag{2.69}$$

Here it is again understood that $\mathbf{f}''$ stands for the solution of (2.62) and correspondingly $\mathbf{f}'''$ is the solution of

$$\mathbf{f}-\mathbf{f}'-\mathbf{f}''' = \mathbf{K}. \tag{2.70}$$

We now assume, to define the heat conduction problem, that

$$N(\mathbf{f},s,x) = N^0(\mathbf{f},s,T)+N^1(\mathbf{f},s), \tag{2.71}$$

where N^0 is the mean phonon number for thermal equilibrium (eq. 2.4), i.e.

$$N^0 = \frac{1}{e^{\beta\hbar\omega}-1}. \qquad (2.72)$$

One easily verifies that

$$\frac{\partial N^0}{\partial T} = N^0(N^0+1)\frac{\hbar\omega}{kT^2}. \qquad (2.73)$$

When inserting (2.71) in (2.69), we can neglect N^1 in the transport term, since this would otherwise give terms of second order in the temperature gradient. In the collision terms one verifies immediately that N^0 by itself gives no contribution (thus confirming that in the absence of a temperature gradient N^0 represents a stationary distribution) and we expect therefore in the collision terms contributions of at least the first order in N^1. Hence N^1 will be proportional to the temperature gradient, and we shall neglect its square. Lastly it is convenient to take out of N^1 a factor similar to that occurring in (2.73),

$$N^1(\mathbf{f}, s) = N^0(\mathbf{f}, s)\{N^0(\mathbf{f}, s)+1\}g(\mathbf{f}, s). \qquad (2.74)$$

Then, using the identity

$$N^0(\mathbf{f}, s)N^0(\mathbf{f}', s')\{N^0(\mathbf{f}'', s'')+1\} = \{N^0(\mathbf{f}, s)+1\}\{N^0(\mathbf{f}', s')+1\}N^0(\mathbf{f}'', s'') \qquad (2.75)$$

if $\qquad \omega(\mathbf{f}, s)+\omega(\mathbf{f}', s') = \omega(\mathbf{f}'', s''),$

and omitting the superscript 0 since now all phonon numbers refer to equilibrium, we have finally the 'Boltzmann equation'

$$N(N+1)\frac{\hbar\omega}{kT^2}\frac{\partial\omega}{\partial f_x}\frac{\partial T}{\partial x} = \frac{\hbar}{32\pi^2\rho}\int d^3\mathbf{f}'\Big\{\sum_{s',s'}|b(\mathbf{f},s;\mathbf{f}',s';s'')|^2 \times$$
$$\times \omega\omega'\omega''\,\delta(\omega+\omega'-\omega'')(N+1)(N'+1)N''(g''-g-g')+$$
$$+\sum_{s',s'''}\tfrac{1}{2}|b(\mathbf{f},s;\mathbf{f}',s';s''')|^2\omega\omega'\omega'''\,\delta(\omega-\omega'-\omega''')\times$$
$$\times(N+1)N'N'''(g'+g'''-g)\Big\} \qquad (2.76)$$

as the condition that the phonon distribution be stationary.

In this form it is obvious that, in the absence of a temperature gradient, the equation admits the solution

$$g(\mathbf{f}, s) = \text{constant}\times\omega(\mathbf{f}, s), \qquad (2.77)$$

since with this value of g the brackets in (2.76) vanish by virtue of the properties of the δ-functions. This has a very simple physical meaning, since it is easy to see that inserting from (2.77) in (2.74) this yields an expression proportional to (2.73) and thus represents simply an infinitesimal change of temperature. Evidently (2.72) must lead to a stationary distribution whatever the temperature. Now in general an inhomogeneous equation of the type (2.76) will be soluble only if the inhomogeneous part is orthogonal to all solutions of the corresponding homogeneous equation. This is all right for the solution (2.77) since this is even in its directional dependence, i.e. it does not change if we replace f_x by $-f_x$, whereas the left-hand side of (2.76) is odd. In other words, the temperature gradient has no tendency either to increase or to decrease the temperature at any given point.

If we were to omit the Umklapp processes, then the homogeneous equation would admit the further solution

$$g(\mathbf{f},s) = \text{constant} \times f_x, \qquad (2.78)$$

and this would not be orthogonal to the left-hand side.

This fact could be made the basis of a method of solving the equation for low temperatures, if the dispersion law was known with sufficient accuracy to determine the solutions of (2.61) and (2.62) in practice. We could then note that at low temperatures Umklapp collisions are rare, most collisions being of the kind which preserves the distribution (2.78). It is therefore reasonable to suppose that the actual distribution is given by (2.78) with a small correction, smaller in the ratio of Umklapp to ordinary processes. Then we can find the factor in (2.78) by multiplying (2.76) by f_x and summing over all $\mathbf{f}$ and s. In other words, we compute the rate of change of J_x. On the right-hand side, then, only the Umklapp processes matter, and we may then neglect the deviation of the distribution from (2.78) as small of higher order. Then the right-hand side is known except for the constant, and the equation thus determines the constant in (2.78) and through it the energy transport. However, by writing all this down we should only confirm the result (2.64) which we have already derived by intuitive arguments; we should not be able to make it more quantitative without a much more extensive study of the dispersion law.

2.6. High temperatures

If we are in the region where classical mechanics is valid, $T > \Theta$, then all the phonon numbers occurring in (2.76) will be large, so that $N+1$ may be replaced by N, and they are all proportional to the temperature. Collecting the temperature factors in the equation, it is then easy to see that, for a given temperature gradient, g must be proportional to T^{-3}. Going back to (2.74) this makes the change in the mean phonon number N^1 proportional to $1/T$, and therefore we expect an energy transport, and thermal conductivity, proportional to $1/T$. This law is usually stated to be in agreement with the behaviour of most crystals at high temperatures.

Its theoretical validity has been called in question by Pomeranchuk (1942). His reasoning starts from the fact, which we have already noted, that long-wave longitudinal phonons (or more precisely phonons with $s = 3$) occur only in collisions in which the other phonons have comparable wavelength. Now consider the Boltzmann equation (2.76) for the particular case $s = 3$, and very small f. Then the left-hand side is proportional to f^{-1}. Since f' is also of order f, the integration over $\mathbf{f}'$ covers a volume of order f^3. However, the δ-function allows only a surface within this volume, which has an area proportional to f^2. The phonon numbers are inversely proportional to the frequencies and therefore cancel the frequency factors. The coefficient b had been found constant for small frequencies. Hence the right-hand side is proportional to f^2 times some average value of g, averaged over arguments of the order f. The equation therefore requires that g be of the order of f^{-3} for small f. This makes N^1 in (2.74) proportional to f^{-5}, and hence the energy transport

$$\frac{V}{(2\pi)^3} \sum_s N^1 \hbar\omega \frac{d\omega}{df} f^2 \, df d\Omega$$

diverges for small f as $\int f^{-2} df$. Hence an exact solution of (2.76) would lead to an infinite thermal conductivity. Pomeranchuk also points out that the long waves will be much more strongly damped by the effect of the quartic terms, which involve four-phonon processes. In that case the collision of a long-wave with a short-wave phonon of any kind of polarization is always possible, and therefore the difficulty does not arise. However, the terms due to four-phonon processes contain an extra power of T, and if they alone were present the thermal conductivity would be proportional to T^{-2}. In actual fact we must assume that the cubic terms in (2.76) are the dominant ones, and that the quartic ones will come in only for the purpose of limiting the movement of the long-wave phonons. The temperature law should therefore be some compromise between $1/T$ and $1/T^2$. Pomeranchuk gives an argument which leads to a $T^{-\frac{4}{3}}$ law, but, since this is based on the concept of mean free path, which is not really applicable in these circumstances, one cannot regard this conclusion as established. The only prediction that would seem to follow with certainty is that the law must be intermediate between T^{-1} and T^{-2}.

This seems a poor return for a long discussion, but progress beyond this stage is difficult, unless we could construct a dispersion law which was simple enough to allow us to list the solutions for the possible phonon collisions explicitly, and yet realistic enough to give the right kind of collisions including those of the Umklapp type.

2.7. Impurities and size effect

So far our discussion has been concerned with ideal crystals. If the lattice is disturbed by impurities or by imperfections in the growth of the crystal, these will also scatter phonons. Since in this case the scattering centres are present no matter what the temperature, the scattering probability of phonons of a given wave vector and polarization is independent of the temperature. At high temperatures, when the spectral distribution of phonons is fixed, and when the specific heat is constant, such imperfections therefore contribute an additional resistance which is independent of the temperature. The situation is more complicated at low temperatures when we are dealing with phonons of longer and longer wavelength. In that case the temperature variation depends on the variation of the scattering cross-section with wavelength. If the imperfections are small, perhaps single misplaced atoms, they will scatter long waves very inefficiently. They may, in fact, not be sufficient by themselves to give a finite resistance because, as in the case discussed in the preceding section, the long-wave phonons retain too great a mobility. Nevertheless, they may have a large effect on the resistance at low temperatures, if they can cause a sufficiently rapid change in the total wave vector J, so as to make the Umklapp processes unnecessary. In such cases the thermal resistance is again the combined result of several different types of collisions, and any discussion of experimental results becomes accordingly very complex (cf., for example, the paper by Klemens (1951), whose use of the concept of relaxation time is not, however, very convincing).

A further point of interest can be seen by comparing the crude formula (2.56) with a result like (2.64). At low temperatures the conductivity increases rapidly, in spite of the decrease in the specific heat. It follows that the mean free path of the phonons must become very large.

For example, the thermal conductivity of pure KCl at $4°$ K. is of the order of 1 cal./sec. deg. cm. Since the specific heat at that temperature is about 10^4 cal./deg. cm.^{3} and the sound velocity about 6.10^5 cm./sec., the effective mean free path must be of the order of 10^{-1} cm. It is evident, therefore, that the size of the crystal actually used and the condition of its surface are of great importance in the interpretation of the experiments. Experiments by de Haas and Biermasz (1935) and their analysis by Casimir (1938) have indeed verified this dependence on the size of the specimen.

Here again the problem is complicated by the fact that, even when the specimen is much smaller than the mean free path estimated from the crude relation (2.56), this does not mean that the collisions between phonons are completely negligible. The reason is that we have found the thermal resistance to be dependent on the rare Umklapp processes. In the circumstances specified we can therefore merely be sure that a phonon has a small probability of undergoing an Umklapp process on crossing the specimen once, but it may make many collisions in which the wave vector is conserved; these by themselves would not be sufficient to produce a finite resistance, but they may affect the phonon distribution. Hence it is in general necessary to discuss the effect of boundary scattering jointly with that of collisions between phonons. An attempt to do this has been made in the paper by Klemens to which I have already referred, but much remains to be done.

Lastly, it should be mentioned that the collisions between phonons which we have discussed in the last few sections are also responsible for the damping of sound waves in perfect crystals. This has been discussed by Slonimsky (1937). It would take us too far, however, to enter into details of this problem. As far as I am aware, there exist no experiments with which to compare this theory.

Note added in proof

The expansion of the potential energy in powers of the displacements might appear questionable in the case of displacements like (2.20), in which atoms far from the origin are necessarily displaced by an amount much larger than the lattice spacing. However, the potential energy is really a function of the distances between the atoms (hence the identities (1.22) and (2.23)), and the expansion is therefore valid provided the distance between any two atoms changes only by a small fraction. This is compatible with (2.20) provided ϵ is a small number.

Note added January 1956

C. Herring (*Phys. Rev.* **95**, 954, 1954) has shown that the frequency curves in a general direction in the lattice are sufficiently different from those shown in Figs. 1-3 to allow in most cases collisions of a long wave longitudinal phonon with other short wave phonons, and hence the arguments of section 2.6 do in fact give a $1/T$ law for most crystals.

III

INTERACTION OF LIGHT WITH NON-CONDUCTING CRYSTALS

3.1. Statement of problem. Infra-red absorption

IF a process of absorption, emission, or scattering of light is described in quantum mechanics, it always involves two states of the mechanical system (in our case the crystal), namely, the initial and the final state. For the purpose of this chapter we shall assume that both these states can be described by the formalism previously developed, i.e. that we need not take explicit account of the motion of the electrons inside the atoms or ions. This is equivalent to saying that all atoms or ions should, before and after the optical process, be in their ground states. In other words, we exclude from our consideration at present the absorption and emission of visible or ultra-violet light, and the absorption and incoherent scattering of X-rays. These will have to be discussed later. Our assumptions do cover, however, the absorption of infra-red radiation, the coherent scattering of light at any frequency, and the most important type of Raman effect.

Starting with the simplest of these problems, the absorption of infrared radiation, we are therefore concerned with a process in which initially we deal with the crystal in the ground state and an incident light quantum, whereas in the final state the quantum energy appears in the form of vibrations or phonons.

This process can take place only if the vibration of the crystal is associated with the motion of electric charges or currents, and in practice this restricts us to ionic crystals. Strictly speaking, even in other crystals the atoms may carry electric or magnetic multipoles either permanently or as a result of their distortion by the surrounding lattice. A vibration of the atoms may therefore produce electromagnetic fields by the changes in these multipole moments; however, in practice this effect is completely negligible.

We assume, therefore, that we are dealing with an ionic crystal. It is then a good approximation to assume that in the course of a vibration the charge on each ion is bodily displaced. This again is not quite rigorous; if an ion moves with respect to its neighbours, the repulsive forces will tend to prevent the movement of the outer electron shell more strongly than that of the inner core, with the result that the displacement of the centre of charge of the ion may exceed that of the centre of mass in the case of a positive ion, and be less in the case of a negative ion. These effects may cause a small correction to the intensity of the absorption or emission lines, and they will not affect the selection rules at all. We shall therefore ignore them.

This means that we may take the whole charge of the ion as concentrated at its centre. Then the interaction energy with an external electromagnetic field of vector potential $\mathbf{A}(\mathbf{r},t)$ and no scalar potential may be written in the form†

$$\sum_{\mathbf{n},j} \left\{ \frac{1}{c} e_j \mathbf{A}(\mathbf{R}_{\mathbf{n},j}) \cdot \dot{\mathbf{R}}_{\mathbf{n},j} + \frac{e_j^2}{2M_{\mathbf{n},j}c^2} \{\mathbf{A}(\mathbf{R}_{\mathbf{n},j})\}^2 \right\}, \qquad (3.1)$$

where e_j is the charge on the jth ion in the $\mathbf{n}$th cell, $\mathbf{R}_{\mathbf{n},j}$ is its position. The second term, which is proportional to the square of the vector potential, has matrix elements only for transitions in which the number of photons changes by two or remains unchanged. It is therefore of interest in connexion with the scattering of light and with Raman effect, but does not contribute to simple absorption or emission.

In the first term we assume, as before, that the ions are nearly in their equilibrium positions. Since $\dot{\mathbf{R}}$ is then a small quantity of the first order, we may, in the argument of $\mathbf{A}$, replace $\mathbf{R}_{\mathbf{n},j}$ by its equilibrium value $\mathbf{a}_\mathbf{n}+\mathbf{d}_j$. The velocity $\dot{\mathbf{R}}$ is, of course, the same as the rate of change of the displacement, $\dot{\mathbf{u}}$, and if we express this in terms of normal coordinates by (1.31) and (1.32) we obtain for the interaction energy

$$-\frac{i}{c} \sum_{\mathbf{n},j} \sum_{\mathbf{f},\sigma} e_j \mathbf{A}(\mathbf{a}_\mathbf{n}+\mathbf{d}_j) \cdot \mathbf{v}_j(\mathbf{f},\sigma) e^{i\mathbf{f}\cdot\mathbf{a}_\mathbf{n}} \omega(\mathbf{f},\sigma) Q(\mathbf{f},\sigma). \qquad (3.2)$$

To find the transition probability induced by this interaction, we use again the standard relation (2.58). $\mathbf{A}$ has matrix elements corresponding to the absorption and emission of one photon. Fixing our attention on an absorption process, for a photon of wave vector $\mathbf{k}$, polarized in the direction of unit vector $\boldsymbol{\epsilon}$, and subject to the same cyclic boundary condition as the vibrations, the matrix element of $\mathbf{A}$ becomes

$$\boldsymbol{\epsilon} \left(\frac{\hbar c^2}{4L^3 \Omega} \right)^{\frac{1}{2}} e^{-i\mathbf{k}\cdot(\mathbf{a}_\mathbf{n}+\mathbf{d}_j)}, \qquad (3.3)$$

where $\Omega = ck$ is the light frequency. Evidently this process can conserve energy only if the vibrational energy increases, i.e. if the phonon number is increased. Hence σ must be positive, and taking the form of the matrix

† See, for example, Schiff (1949), section 23.

element from (1.63) we have finally for the matrix element for conversion of the photon of wave vector **k** and polarization **ε** into a phonon of wave vector **f** and polarization s

$$-i\left(\frac{\hbar}{4L^3\Omega}\right)^{\frac{1}{2}}\left(\frac{\hbar\omega(\mathbf{f},s)}{2M^{(N)}}\right)^{\frac{1}{2}}\sum_j e^{-i\mathbf{k}\cdot\mathbf{d}_j}e_j\,\boldsymbol{\epsilon}\cdot\mathbf{v}_j(\mathbf{f},s)\sum_\mathbf{n} e^{i(\mathbf{k}-\mathbf{f})\cdot\mathbf{a}_\mathbf{n}}. \qquad (3.4)$$

The last factor is again a sum of the familiar form (1.40) and vanishes unless **k**−**f** is a lattice vector in the reciprocal lattice. When this is the case each term in the sum equals unity, and the sum reduces to the number N of unit cells in the crystal. Hence the transition probability becomes

$$\delta\{\omega(\mathbf{f},s)-\Omega\}\frac{\pi N^2}{4L^3 M^{(N)}}\left|\sum_j e_j\,\boldsymbol{\epsilon}\cdot\mathbf{v}_j(\mathbf{f},s)e^{-i\mathbf{k}\cdot\mathbf{d}_j}\right|^2, \qquad (3.5)$$

where $$\mathbf{f}-\mathbf{k} = \mathbf{K}, \qquad (3.6)$$

while the δ function singles out the value of **k** for which

$$\omega(\mathbf{f},s) = ck. \qquad (3.7)$$

The conditions (3.6), (3.7) are illustrated in Fig. 4. The full lines represent the various branches of the vibration spectrum, the broken line gives the frequency of the photon, and any intersection is a solution of (3.6). The figure must, of course, be regarded as a one-dimensional section of a three-dimensional problem.

To be realistic the broken line should be much steeper than is shown, since its slope represents the velocity of light, whereas the initial slope of the full lines is the velocity of sound, which is about $10^{-5}c$. It therefore follows at once that there is no intersection for those branches whose frequency vanishes when $\mathbf{f} = 0$, which we called the acoustic branches. Moreover, the value of **f** at the intersection is extremely small, of the order of the ratio of sound to light velocity, and may, for most purposes, be replaced by zero. Lastly it is evident that the reciprocal lattice vector **K** in (3.6) will, in fact, be zero.

Since k must be very small compared to the inverse lattice spacing, it follows that $\mathbf{k}\cdot\mathbf{d}_j$ is a very small quantity and can usually be neglected. The factor $N^2/L^3 M^{(N)}$ can be written in the form $\rho/(M^c)^2$, where ρ is again the density and $M^{(c)}$ the mass of a unit cell. So finally (3.5) becomes

$$\delta\{\Omega-\omega(0,s)\}\frac{\pi\rho}{4(M^c)^2}\left|\sum_j e_j\,\boldsymbol{\epsilon}\cdot\mathbf{v}_j(0,s)\right|^2. \qquad (3.8)$$

The simplest case is that of an NaCl type lattice, $r = 2$. In this case there are three acoustic branches and three 'optical' ones. For $\mathbf{f} = 0$,

§ 3.1 NON-CONDUCTING CRYSTALS 57

the displacement of all positive ions must be the same, and similarly all negative ions must have the same displacement. Moreover, from the orthogonality with the acoustic vibration of infinite wavelength for which all displacements are equal, we can deduce from (1.41) or (1.42) that the positive and negative ions will be displaced in opposite directions by amounts in the inverse ratio of their masses.

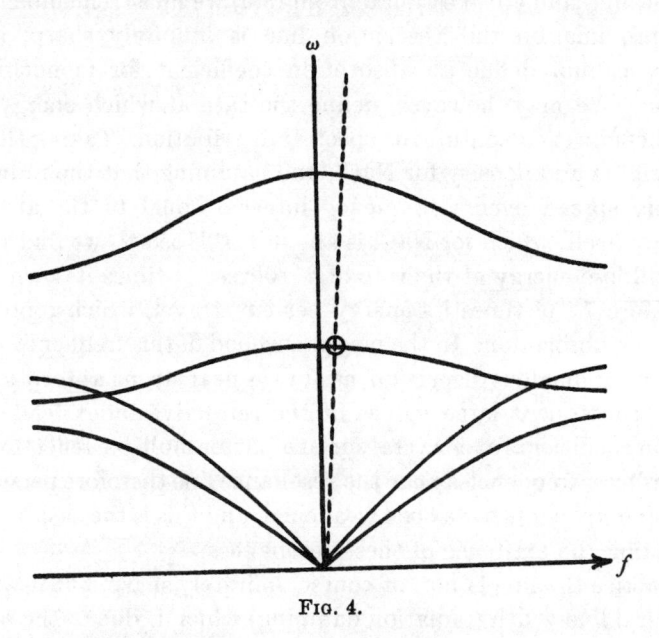

Fig. 4.

For the three optical branches, $s = 4, 5$, and 6, the displacements of the positive and negative ions are in opposite directions. In this case, the frequency depends on this direction even when $\mathbf{f}$ is very small, because of the long-range effect of electrostatic forces (cf. Kellerman, 1940). If f is small, but the wavelength $1/f$ is small compared with the dimensions of the crystal, one normal mode is exactly longitudinal and the others transverse; the frequency of the longitudinal one for small f is larger than that of the transverse ones. This has been indicated in Fig. 4. For the longitudinal mode $\mathbf{v}_j$ is parallel to $\mathbf{f}$, i.e. perpendicular to $\boldsymbol{\epsilon}$. For this mode we see that the transition probability (3.8) vanishes. There is therefore only one absorption frequency, corresponding to the marked intersection in Fig. 4.

In a cubic crystal the transverse modes must, for small f, have the same frequency, and because of this we may choose one of them to be polarized in the direction of $\boldsymbol{\epsilon}$. For this mode the transition probability becomes

(using the orthogonality and normalization)

$$\delta\{\Omega - \omega(0, 4)\} \frac{\pi \rho e_1^2}{M_1 M_2}, \qquad (3.9)$$

where M_1, M_2 are the masses of the two ions, and $e_1 = -e_2$ the charge of one of them.

To obtain a numerical estimate from (3.9) we must remember that to this approximation the absorption line is infinitely sharp, and we therefore cannot define an absorption coefficient for monochromatic radiation. We may, however, define the rate at which energy is lost from a beam with a continuous spectral distribution. Taking the molecular weights and density for NaCl, and assuming that the radiation is uniformly spread over a frequency interval equal to the absorption frequency itself, which for NaCl is about 4.10^{13} sec.$^{-1}$, we find that the beam will lose energy at the rate of 2.10^{15} sec.$^{-1}$ times its own energy, or therefore 7.10^4 times its energy per cm. travel, which represents a very strong absorption. In the neighbourhood of this frequency one will find strong anomalous dispersion, as always near strong absorption lines, and the consequent large values of the refractive index lead to high reflection coefficients at a crystal surface. After multiple reflections from such surfaces, frequencies near the resonance line therefore persist while all other components have been weakened, and this is the easiest method of detecting the existence of these resonances.

In practice the line is not, of course, infinitely sharp, but it will show the natural line width (radiation damping) which is due to the fact that the re-emission of the radiation which has been absorbed limits the lifetime of the excited state. In addition, the excited state may, by the anharmonic coupling terms, transform into a state in which two different phonons replace the optical one, and this process also limits the life of the phonon in the optical branch and hence causes a broadening.

Finally, it should be remarked that all the results of this section could have been obtained classically (which might be guessed from the fact that the final result (3.8) does not contain the quantum constant). We have, nevertheless, chosen to use quantum mechanics both because the calculation then links up more directly with that of other phenomena, and also because it allows us to use concepts like phonon and photon, which help in visualizing the process.

3.2. Diffraction of X-rays

Turning now to the scattering of light, we begin with the simplest case, in which we may neglect the displacement of the atoms from their equilibrium positions, and assume the lattice to be perfect. We also restrict ourselves to the case of coherent scattering.

In that case the amplitude scattered by an atom can be expressed as

$$A_{sc} = -F \frac{e^2}{mc^2} \frac{e^{ikR}}{R} (\boldsymbol{\epsilon}_{in} \cdot \boldsymbol{\epsilon}_{sc}) A_{in}. \tag{3.10}$$

Here R is the distance from the atom at which the scattered wave is observed; R is assumed large compared with the wavelength. $\boldsymbol{\epsilon}_{in}$ and $\boldsymbol{\epsilon}_{sc}$ are the polarization vectors of the incident and scattered waves. A_{in} is the vector potential of the incident wave at the centre of the atom. F is the so-called 'form factor', a number which is in general complex, and which depends on the nature of the atom, the wavelength of the light, and the angle of scattering. The dimensional constant $-e^2/mc^2$ has been introduced so as to make $F = 1$ in the case of a single electron scattering classically, and e, m are the charge and mass of the electron, c is the velocity of light. The equation is, of course, to be understood in the usual sense in which a real periodic potential is represented as the real part of a complex quantity.

This equation assumes the atom to be spherically symmetric; otherwise F would depend on the incident and scattered directions separately. This case is not, however, of practical importance.

Consider now the extension of the above result to the case when the atom is not at the origin, but at some point $\mathbf{r}$. In that case the incident amplitude compared with that at the origin contains the factor $e^{i\mathbf{k}\cdot\mathbf{r}}$, where $\mathbf{k}$ is the wave vector of the incident wave. Similarly, the scattered amplitude at a point a long distance R from the origin will differ from that at a point distant R from the atom by a factor $e^{i\mathbf{k}'\cdot\mathbf{r}}$, where $\mathbf{k}'$ is the wave vector of the scattered wave. This means we must multiply (3.10) by

$$e^{i(\mathbf{k}-\mathbf{k}')\cdot\mathbf{r}}. \tag{3.11}$$

Now consider the whole crystal. We assume for simplicity that the scattering is weak, and that therefore the wave incident on each atom is the original incident wave, and that the waves scattered by other atoms may be neglected in comparison with it. Then we obtain the total scattered amplitude by adding for each atom a term (3.10) multiplied by a factor (3.11) to take account of its position. For the perfect lattice therefore

$$A_{sc} = -\frac{e^2}{mc^2} \frac{e^{ikR}}{R} (\boldsymbol{\epsilon}_{in} \cdot \boldsymbol{\epsilon}_{sc}) A_{in} \sum_j \sum_n F_j e^{i\mathbf{q}\cdot(\mathbf{a}_n+\mathbf{d}_j)}, \tag{3.12}$$

where

$$\mathbf{q} = \mathbf{k} - \mathbf{k}'. \tag{3.13}$$

The summation over **n** then gives rise to the familiar factor

$$\sum_{\mathbf{n}} e^{i\mathbf{q}\cdot\mathbf{a_n}}, \quad (3.14)$$

which vanishes unless very nearly

$$\mathbf{q} = \mathbf{K}, \quad (3.15)$$

where **K** is a reciprocal lattice vector.

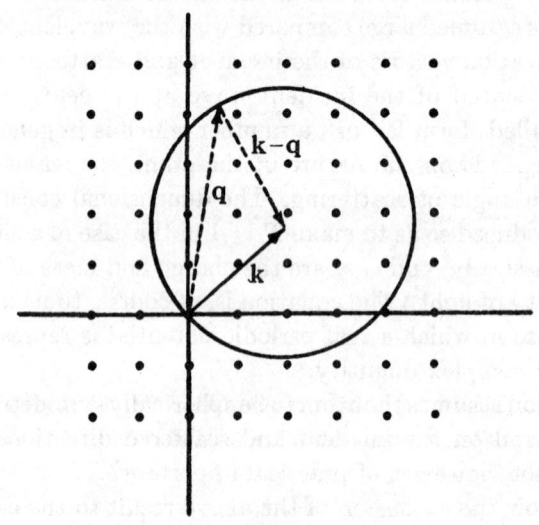

Fig. 5.

Since we are concerned with coherent scattering, the vectors **k** and **k'** must be equal in magnitude. Hence each must be at least one-half the magnitude of the smallest non-zero lattice vector of the reciprocal lattice. (The case when $\mathbf{K} = 0$ means no scattering.) Since the smallest **K** is of the order of $2\pi/a$, where a is a length of the order of 10^{-8} cm., we see therefore that the perfect lattice does not scatter light coherently unless the wavelength is in the X-ray region or shorter.

In general, for a given incident wave and a given orientation of the crystal, no scattering is possible. Since **k** is given, and **k'** must have the same magnitude, we may construct all possible vectors **k'** by drawing, in reciprocal space, the sphere

$$|\mathbf{k}-\mathbf{q}|^2 - k^2 = 0,$$

or
$$q^2 - 2\mathbf{k}\cdot\mathbf{q} = 0 \quad (3.16)$$

which passes through the origin (cf. Fig. 5). In general, none of the values of **q** which belong to points on this sphere will coincide with any of the points in the reciprocal lattice.

In practice, the scattering of X-rays is observed by one of two methods. In the first the X-rays used are not monochromatic (Laue method). Then we have to replace the sphere by a family of spheres obtained by varying the magnitude (but not the direction) of k. One of the spheres will then pass through each reciprocal lattice point, defining a given vector k'. Each lattice point therefore results in scattering in a particular direction, and a photographic plate will record a network of points, each belonging to one point of the reciprocal lattice. The wavelength of the radiation scattered by each point is then different, and this makes a quantitative use of this method rather inconvenient.

Alternatively, one uses monochromatic radiation, but varies the direction of the crystal. This can be done by rotating a single crystal about a suitable axis. The effect of this is to keep the sphere of Fig. 5 fixed, but rotate the lattice. It is evident that in the course of the rotation some lattice points will cross the sphere, thus making scattering possible. A similar result is achieved by using, instead of a single crystal, a piece of polycrystalline material or a powder made up of small crystals with random orientations. This is equivalent to rotating the lattice of Fig. 5 into all possible orientations. In that case each lattice vector K will yield scattered radiation covering a cone with the incident direction as axis (Debye-Scherrer method).

Provided the crystal is oriented suitably with respect to the incident beam to satisfy the diffraction condition (3.16) for a particular vector K, we have for the magnitudes

$$2k \sin \tfrac{1}{2}\theta = K,$$

where θ is the angle between incident and scattered radiation; or, remembering the connexion between wave vector and wavelength,

$$\frac{2}{\lambda} \sin \frac{\theta}{2} = \frac{K}{2\pi}. \qquad (3.17)$$

The right-hand side is an inverse length which can be shown to be equal to a multiple of the inverse distance between suitable planes containing atoms in the original lattice. In the form

$$\frac{2}{\lambda} \sin \frac{\theta}{2} = \frac{n}{d} \qquad (3.18)$$

the relation is known as Bragg's law.

The intensity of the reflection belonging to a particular order, i.e. to a particular lattice vector K, depends, according to (3.12), on the quantity

$$\sum F_j e^{i\mathbf{K}.\mathbf{d}_j} \qquad (3.19)$$

taken over the unit cell. If the atomic form factors F_j are known for all atoms or ions of the crystal, a measurement of a sufficient number of lines will then determine the $\mathbf{d}_j$ and hence the positions of the atoms in the unit cell.

The form factor F_j may be determined empirically, or from atomic theory. The theoretical determination is in general complicated, but simplifies greatly if the frequency of the X-ray is much greater than the frequencies of the important absorption lines of the atom or ion. One can then prove that in the interaction of the radiation with the atom, which is again given by (3.1) but with e_j, M_j replaced by e, m, the charge and mass of the electron, the first term is negligible compared with the second. Since this term is quadratic in the vector potential, it contains directly matrix elements which represent the disappearance of one photon of wave vector $\mathbf{k}$ and the emission of one of wave vector $\mathbf{k}'$. Using the matrix elements for $\mathbf{A}$ which we have already used in (3.3), we find for the relevant matrix element of the second term of (3.1)

$$\frac{e^2}{2mc^2} \frac{\hbar c^2}{4L^3\omega} (\boldsymbol{\epsilon}_{\text{in}} \cdot \boldsymbol{\epsilon}_{\text{sc}}) \sum_{l=1}^{Z} e^{i\mathbf{q}\cdot\mathbf{r}_l}, \tag{3.20}$$

where the sum goes over the electrons in the atom, and $\mathbf{r}_l$ is the position of the lth electron. Of this expression we should take now the matrix element between appropriate atomic states. Since in coherent scattering the initial and the final state are the same, this amounts simply to averaging (3.20) over the electron coordinates with the density function describing the distribution of electrons in the atom.

Making the usual transition from matrix element to scattering amplitude, one obtains for F the expression

$$\int \rho(\mathbf{r}) e^{i\mathbf{q}\cdot\mathbf{r}} d^3\mathbf{r}, \tag{3.21}$$

where $\rho(\mathbf{r})$ is the electron density at the point $\mathbf{r}$. In this case, the total form factor of the unit cell may be obtained by inserting (3.21) in (3.19), with the result

$$\int \sum_j \rho_j(\mathbf{r}) e^{i\mathbf{q}\cdot\mathbf{r}} d^3\mathbf{r}. \tag{3.22}$$

Here $\mathbf{r}$ is counted from a fixed origin in the unit cell, and the integral goes over the volume of the unit cell. $\sum_j \rho_j(\mathbf{r})$ is the sum of the electron densities due to all atoms, and hence just the total electron density. It is therefore not necessary to decide just which part of the electron distribution belongs to one or the other atom. This might in fact be rather difficult for the outer parts of the atoms, which overlap.

(3.22) shows, in fact, that the observations give us directly the Fourier coefficients of the electron density in the unit cell.

As I have pointed out, these formulae are valid only if the atomic resonance frequencies can be neglected in comparison with the X-ray frequency. The most important example of the opposite is the case when the X-ray is near the absorption edge of one of the shells of an atom in the crystal. In that case that particular shell may make a large contribution to the scattering, because of anomalous dispersion. In such cases equations (3.20) to (3.22) fail. They must also be corrected for very hard X-rays and γ-rays, when relativistic corrections become important.

According to the simple theory of this section, the X-ray reflections are infinitely sharp, since (3.15) must hold exactly. In fact, even for an ideal crystal, the lines have a finite width, because, if a wave has suitable wavelength and frequency to be scattered, it will be weakened rapidly in passing through the crystal, and instead of using plane sinusoidal waves in (3.12) one should really give them an exponentially diminishing amplitude. In that case the condition (3.15) is no longer rigorous, but only approximate. A quantitative theory of this problem is contained in Ewald's 'dynamical theory' of X-ray interferences. The problem is closely related to the width of the lines in electron diffraction, which we shall have to discuss from rather a different point of view.

3.3. Effect of the atomic vibrations

In the real crystal the atoms are subject to displacements from the ideal lattice positions, as we saw in Chapter I. We must therefore correct (3.12) by substituting the real position of each atom for the lattice point. In this way, the scattered amplitude will depend on the displacements, which are subject to statistical fluctuations, and also vary with time. Hence the scattered amplitude will no longer have exactly the same frequency as the incident radiation, and this should, strictly speaking, be allowed for by correcting in each case the wave vector $\mathbf{k}'$ of the scattered wave so that its magnitude corresponds to the actual frequency. However, this change of frequency is of the order of the frequencies of the lattice vibrations, and as long as we restrict ourselves to the X-ray region it is quite negligible.

If we correct the sum in (3.12) for the atomic displacements, we have

$$\sum_{\mathbf{n},j} F_j\, e^{i\mathbf{q}\cdot(\mathbf{a}_\mathbf{n}+\mathbf{d}_j+\mathbf{u}_{\mathbf{n},j})}. \tag{3.23}$$

Consider for the moment the region of low-order diffraction, for which $\mathbf{q}-\mathbf{q}'$ is of the order of an inverse atomic distance. Since we have seen that the atomic displacements are small compared with the atomic distance, the term containing **u** in the exponent is then a small number, and we may expand its contribution in a series. Up to second order the result is

$$\sum_{n,j} F_j e^{i\mathbf{q}\cdot(\mathbf{a_n}+\mathbf{d}_j)}\{1+i\mathbf{q}\cdot\mathbf{u}_{n,j}-\tfrac{1}{2}(\mathbf{q}\cdot\mathbf{u}_{n,j})^2\}. \tag{3.24}$$

We shall discuss separately the effects of the first-order and of the second-order terms.

In the first-order term we express the **u** in terms of normal coordinates by (1.30) and (1.32). The average value of each of the Q vanishes, and the only non-vanishing matrix elements are those in which $N(\mathbf{f},s)$ increases or decreases by one unit. This corresponds to a process in which a phonon in this mode is emitted or absorbed. Explicitly the first-order term is

$$\sum_{j}\sum_{\mathbf{f},s} F_j e^{i\mathbf{q}\cdot\mathbf{d}_j} \sum_{\mathbf{n}} e^{i(\mathbf{q}+\mathbf{f})\cdot\mathbf{a_n}} \mathbf{q}\cdot\mathbf{v}_j(\mathbf{f},s)\{Q(\mathbf{f},s)+Q(\mathbf{f},-s)\}. \tag{3.25}$$

It is clear that this term leads to scattering whatever the wavelength of the X-rays, and for any angle of scattering, since for any given vector **q** it is always possible to find an **f** for which the sum over **n** gives a large contribution. This requires that

$$\mathbf{f} = \mathbf{K}-\mathbf{q}, \tag{3.26}$$

where **K** is a vector in the reciprocal lattice. Since **f** was defined to be in the basic cell of the reciprocal lattice, the relation (3.26) determines **f** and **K** uniquely. (3.25) therefore gives a continuous background to the scattering, though, as we shall see, this background is by no means uniform.

To discuss the background intensity, we simplify the problem by assuming that there is only one atom in the unit cell, so that the suffix j may be omitted.

The intensity is proportional to the squared modulus of (3.25), averaged over time, which is equivalent to taking the diagonal matrix element for the initial state of the crystal. Since $Q(\mathbf{f},s)$ represents the absorption of a certain phonon, $Q(-\mathbf{f},-s)$ its emission, only their product will contribute to the diagonal element. (This is the same as saying that processes in which different phonons are produced or absorbed are incoherent with each other.) Hence we have

$$|F|^2 \sum_{n,n'} \sum_{\mathbf{f},s} e^{i(\mathbf{q}+\mathbf{f})\cdot(\mathbf{a_n}-\mathbf{a_{n'}})}|\mathbf{q}\cdot\mathbf{v}(\mathbf{f},s)|^2\{Q(\mathbf{f},s)Q^*(\mathbf{f},s)+Q^*(\mathbf{f},s)Q(\mathbf{f},s)\} \tag{3.27}$$

or, inserting for the matrix elements from (1.63),

$$\frac{|F|^2\hbar}{M^{(N)}} \sum_{n,n'} \sum_{\mathbf{f},s} e^{i(\mathbf{q}+\mathbf{f})\cdot(\mathbf{a_n}-\mathbf{a_{n'}})} \frac{|\mathbf{q}\cdot\mathbf{v}(\mathbf{f},s)|^2}{\omega(\mathbf{f},s)}\{N(\mathbf{f},s)+\tfrac{1}{2}\}. \tag{3.28}$$

As a function of **q**, this expression will show a variation in detail according to whether or not the vector **f** defined by (3.26) coincides with one of the discrete values permitted by the boundary conditions for the vibrations, as expressed by (1.27). This effect is due to the diffraction by the external shape of the crystal, and in practice neither the wavelength nor the direction of the X-rays will be well enough defined to make this observable. We are therefore interested in the average of (3.28) over a region containing many solutions of (1.27), and then the variation with **q** will be due only to the last two factors, in which we may regard **f** as a function of **q** by (3.26). Inserting for the phonon number $N(\mathbf{f}, s)$ its thermal average from (2.2), the background intensity is thus seen to be proportional to

$$\sum_s |\mathbf{q} \cdot \mathbf{v}(\mathbf{f}, s)|^2 \frac{1}{\omega(\mathbf{f}, s)} \coth \frac{\hbar \omega(\mathbf{f}, s)}{2kT}. \tag{3.29}$$

Here the first factor varies very slowly with **f**. The last two factors are inversely proportional to $\omega(\mathbf{f}, s)^2$ at high temperatures, when the coth may be replaced by the inverse of its argument. In that case the function will be greatest near $\mathbf{f} = 0$, and, because of (1.63), will there vary as $1/f^2$. The background intensity is therefore inversely proportional to the square of the deviation in reciprocal space, from the nearest reciprocal lattice point. The factor of proportionality depends on the direction of this deviation.

At these high temperatures, the background intensity is proportional to T.

At $T = 0$, when (3.29) is equivalent to putting $N = 0$ in (3.28), so that only the emission but not the absorption of phonons contributes, the intensity varies only as f^{-1}.

Next consider the second-order term in (3.24). Since we have assumed $\mathbf{q} \cdot \mathbf{u}_{\mathbf{n},j}$ to be a small number, its contribution to the amplitude is small compared to that of the first-order term which we have just discussed. It does, however, contain terms which are coherent with the leading term, and these will give contributions to the intensity of the same order as the first-order terms. These coherent parts are those which do not involve any change in the phonon numbers, and therefore represent the average of (3.24) over the initial state of the crystal. The second-order contribution thus becomes

$$-\tfrac{1}{2} \sum_{\mathbf{n},j} F_j e^{i\mathbf{q} \cdot (\mathbf{a_n} + \mathbf{d}_j)} \overline{(\mathbf{q} \cdot \mathbf{u}_{\mathbf{n},j})^2}, \tag{3.30}$$

where the line over the symbol again denotes the average. The last factor depends on the mean square displacement of the atom $\mathbf{n}, j$ in the direction of $\mathbf{q}$. Because of the translational symmetry this must be the same for corresponding atoms in different cells, and hence independent of $\mathbf{n}$. The sum over $\mathbf{n}$ is therefore the same as in the leading term of (3.24). The effect of the correction (3.30) is merely to replace F_j in the formula for the perfect lattice by

$$F_j\{1-\tfrac{1}{2}\overline{(\mathbf{q}\cdot\mathbf{u}_{n,j})^2}\}. \tag{3.31}$$

This is therefore always a reduction of the scattering power, and it amounts to replacing the electron density distribution of the atom by a smaller density, spread over a larger volume, as is appropriate to an atom whose position is subject to fluctuations.

To evaluate the average, we proceed as in the first-order term

$$\overline{(\mathbf{q}\cdot\mathbf{u}_{n,j})^2} = \sum_{\mathbf{f},s} |\mathbf{q}\cdot\mathbf{v}_j(\mathbf{f},s)|^2 \overline{\{Q(\mathbf{f},s)Q^*(\mathbf{f},s)+Q^*(\mathbf{f},s)Q(\mathbf{f},s)\}}$$

$$= \sum_{\mathbf{f},s} |\mathbf{q}\cdot\mathbf{v}_j(\mathbf{f},s)|^2 \frac{\hbar}{2M^{(N)}\omega(\mathbf{f},s)} \coth\frac{\hbar\omega(\mathbf{f},s)}{2kT}. \tag{3.32}$$

For the case of one atom in the unit cell this gives again the last factor (3.27), summed over all $\mathbf{f}$, and in that case one sees therefore that the loss of intensity from the regular Bragg diffraction is just equal to the total intensity in the background (except for the variation of F with $\mathbf{q}$). The situation is not so simple in the case of a more complex unit cell. For example, it may happen that for a certain regular diffraction two atoms contribute with equal and opposite amplitudes and hence leave only a small difference. In that case, if one of them vibrates with greater amplitude than the other, the effect of the vibration may actually be to increase that particular peak.

If there is only one atom in the unit cell, the expression (3.32) may be further simplified if the lattice has cubic symmetry. In that case we are dealing with a quadratic form in the components of $\mathbf{q}$, which must have cubic symmetry, and any quadratic form of cubic symmetry is isotropic. It must therefore be proportional to q^2, and hence we may replace the first factor by $\tfrac{1}{3}q^2|\mathbf{v}(\mathbf{f},s)|^2$,

which, by (1.40), is simply $\tfrac{1}{3}q^2$. Replacing the summation over $\mathbf{f}$ by an integration,

$$\overline{|q\cdot\mathbf{u_n}|^2} = \sum_s \int d\Omega \int f^2\,df \frac{\hbar}{2(2\pi)^3\rho\omega(\mathbf{f},s)} \tfrac{1}{3}q^2 \coth\frac{\hbar\omega(\mathbf{f},s)}{2kT}. \tag{3.33}$$

This expression is again proportional to T at high temperatures, and reaches a constant limit at low temperatures, the first increase above this limit being proportional to T^2, as may be seen from the same reasoning that led to (2.12).

The behaviour of the integral near $f = 0$ is of some interest. In that region the frequency is proportional to f (the factor depending on direction and polarization) and, since the coth goes as the inverse argument for small argument, the integrand is constant. If, however, we were dealing with a two-dimensional lattice, or with a linear chain, the integrand would go as f^{-1} or f^{-2} respectively, and the integral would diverge. This means that we should then not be justified in replacing the sum by an integral. It is easy to see that the sum would increase with the logarithm of the linear dimensions in the plane case, and proportionally to the length in the case of the chain.

The reason for this is particularly evident in the case of the linear chain if we assume that each atom interacts only with its nearest neighbours. In that case, if the position of the left-hand end of the chain is held fixed, a small displacement of any one atom will shift the equilibrium position of all other atoms to the right of it, and there is no mechanism for correcting this error again later. Hence all errors in the spacing are cumulative for the absolute position of the atoms, and one must expect that the mean square displacement should increase linearly with distance, as our calculation indicates. If there are forces between atoms that are not immediate neighbours, the argument is less obvious, but it is still true that any small displacement of a group of atoms will destroy the exact alignment of the atoms on either side of it.

In the three-dimensional case, on the other hand, any two atoms can be linked by so many different paths that they have a much stronger tendency to maintain their relative positions. This so-called 'long-distance' order is thus characteristic of the three-dimensional case.

Equation (3.31) can easily be generalized to the case of larger $\mathbf{q}$ (high-order diffraction) when $\mathbf{q}\cdot\mathbf{u}$ is no longer small. This is because the probability distribution of the displacement of an atom is given by an error function. This is evident at high temperatures, where it is an immediate consequence of Boltzmann's law, and at zero temperature, when it follows from the fact that the wave function of a harmonic oscillator in its lowest state is an error function. It can, however, also be shown for the intermediate case (e.g. Born, 1943). Therefore the mean value of (3.23) can be obtained exactly, and turns out to be an exponential of which (3.31) is the beginning of the expansion. In other words, the more exact form of (3.31) is

$$F_j e^{-\frac{1}{2}|\mathbf{q}\cdot\mathbf{u}_{n,j}|^2}. \tag{3.34}$$

The exponent is still given by (3.32).

For such large values of **q** the expression for the background must also be modified, but this problem is less simple, and also less interesting. Since one is now concerned with the simultaneous emission or absorption of several phonons, one can see that the effect will be to reduce the maximum of the background near the Bragg diffraction, and spread it out more widely.

3.4. Scattering of light

We have seen in the last section that, whereas scattering of radiation by a rigid lattice is possible only for waves in the X-ray region or shorter, the atomic vibrations effectively remove this limitation. We may therefore apply the method of the last section also to the scattering of longer waves, for example of visible light.

In that region some of the approximations made in deriving our basic equations need revision, but it is easiest to consider first what happens if we apply (3.23) to the visible region, and discuss corrections later.

In the visible region the wavelength is very large compared to the lattice constant, and therefore k and k' are small. In this case $\mathbf{q} \cdot \mathbf{u}$ in (3.24) is certainly small so that we need not go beyond the first order, particularly since there is no solution to the Bragg conditions, as we have seen.

Hence (3.25) then contains the entire effect, and again the only contribution comes from values of **f** for which (3.26) is valid, which in this case requires that $\mathbf{K} = 0$, $\mathbf{f} = -\mathbf{q}$.

For a lattice with one atom per unit cell, a wave with the small wave vector $-\mathbf{q}$ is an elastic wave; and in this case a knowledge of the elastic constants is in principle sufficient to describe the scattering.

If in particular **q** has the direction of one of the axes of symmetry of the crystal, there will be three normal modes, of which one is longitudinal, the other two transverse. The product $\mathbf{q} \cdot \mathbf{v}(-\mathbf{q}, s)$ vanishes for the transverse waves, so that then only one of the three values of s contributes. For a general direction **q** the three directions of polarization will deviate from the transverse and longitudinal directions by an amount dependent on the anisotropy of the elastic constants, and then in general all three will contribute to the scattering, though in different amounts.

In the case of visible light the change in frequency of the light due to the absorption or emission of a phonon is observable. It is, however, still a small change, so that we can find **q** as if **k** and **k'** were equal in magnitude, which gives

$$q = 2k \sin \tfrac{1}{2}\theta, \tag{3.35}$$

where θ is again the angle of scattering. Then, the energy of the scattered photon must equal that of the incident photon plus or minus the phonon energy, or

$$\frac{\Delta k}{k} = \frac{k'-k}{k} = \pm \frac{c_s}{c_L} 2\sin\tfrac{1}{2}\theta, \tag{3.36}$$

where c_L is the velocity of light and c_s the sound velocity for waves of polarization s travelling in the direction of $\mathbf{q}$. This ratio is of the order of 10^{-5}.

In a crystal whose elastic constants were isotropic we should thus see one line in the spectrum of the scattered radiation on each side of the original frequency. That with decreased frequency corresponds to the emission of a phonon ('Stokes' line) and that with increased frequency arises from the absorption of a phonon ('anti-Stokes' line). The names in brackets refer to Stokes's rule that in fluorescence the frequency of the emitted radiation is less than that of the incident radiation. The rule is valid for systems in their ground states and it is evident that at very low temperature the anti-Stokes line would disappear, although such temperatures are not easy to reach in practice.

In a real crystal we expect, for a general direction of scattering, to have three Stokes and three anti-Stokes lines; that belonging to the most longitudinal mode, which in general has the greatest sound velocity, should be much stronger than the others.

Let us now return to the question of the justification for (3.24). This was based on the picture that each atom scattered light waves independently, and that, upon a displacement of the atom, it still scattered in the same way as before. However, a large contribution to the scattering in the visible region is due to the outer electrons, which may, in fact, be shared between adjacent atoms, and these outer shells of the atom in any case will not move bodily if the atom is displaced with respect to its neighbours.

However, we are concerned only with the effects of very long-wave vibrations. In these the relative displacement of adjacent atoms is very small, and the main effect is a fluctuation in the density. Indeed the relative change in density is

$$\frac{\partial}{\partial \mathbf{a_n}} \mathbf{u_n},$$

which for a wave $\mathbf{f}$, s of unit amplitude is $e^{i\mathbf{f}.\mathbf{a_n}} i\mathbf{f}.\mathbf{v}(\mathbf{f},s)$ in agreement with (3.25).

Hence all that has, in fact, gone into our use of (3.25) in this case is the fact that, if a region of the crystal is uniformly compressed, the scattering of light per unit volume increases, to first order, as the density. This is a much weaker assumption than was necessary for deriving (3.25) in the case of X-rays, and is justified even in the optical region. (3.25) would not give correctly the scattering of visible light by very short lattice waves, but since this scattering is negligible this does not matter. These results were first described by Brillouin (1922).

Now turn to a crystal with a complex unit cell. Then the sum over s in (3.25) covers not only the 'acoustic branch' of the vibration spectrum, for which the frequency vanishes for $\mathbf{f} = 0$, but besides this, $3r-3$ other modes with finite frequency. We may here neglect $\mathbf{q}$ both in the frequency and in the polarization vector $\mathbf{v}_j(-\mathbf{q}, s)$. We are then dealing with vibrations in which the displacement of corresponding atoms in different cells is the same. In a lattice with cubic symmetry three such modes have the same frequency since the frequency must remain the same if the unit cell is rotated through 90° about one of the cube axes. In that case there are only $r-1$ distinct frequencies.

We then find that the scattered light contains lines shifted in frequency to either side by the frequencies of the internal vibrations of the unit cell. In this case the order of magnitude of the shift is some 10^{-2} eV. This phenomenon, which is one aspect of the Raman effect, is very useful in investigating the geometrical and dynamical properties of the unit cell.

For this effect it would not be correct to calculate the amplitude from (3.25), since in the relative displacement of different atoms in the unit cell we must not expect each atom to move like a rigid body, particularly as regards its outer shells. Our conclusion about the relation between the wave vectors of the initial and scattered photon, and the absorbed or emitted phonon, and between their energies, are still valid, since they are based only on symmetry considerations. An estimate of the intensity of the Raman lines, would, however, need a different approach. (See G. Placzek (1934).)

3.5. Scattering of neutrons[†]

It is convenient to discuss in this context the scattering of neutrons by crystals, because in spite of the physical differences the mathematical problem is very similar to those that we have just discussed.

We may assume that the neutrons interact only with the atomic nuclei in the crystal. Their interaction with electrons is very weak, except through the magnetic moment of the neutron interacting with the magnetic field of the electron. In a substance which is chemically saturated there are no magnetic fields. Even if the atoms have magnetic

---

[†] For a more quantitative treatment, see Placzek (1952).

moments, these will be oriented at random and at most give background scattering. The special case of ferromagnetic crystals will be discussed later.

For the scattering of neutrons by an atomic nucleus the ratio of scattered to incident amplitude is similar to (3.10)

$$A_{sc} = S \frac{e^{ikR}}{R} A_{in}, \qquad (3.37)$$

where S is now the 'scattering length' and is independent of the angle of scattering and, over a wide range, of the wavelength. The situation is complicated by the fact that the neutron possesses a spin, which is capable of two orientations, and the nucleus may also have a spin. If so, two different processes are possible; either the components of both spins in some direction are not changed by the collision, or both are changed by the collision. In the latter case, the nucleus is no longer in its initial state, and therefore the processes of scattering by different nuclei lead to different final states of the whole system, and must be treated as incoherent. For the purposes of diffraction we should therefore include in (3.37) only the scattering without spin change, and the other process will add a continuous background to the scattered intensity.

The scattering length S in (3.37) will still depend on the initial orientation of the nuclear and neutron spins, and will therefore be different for different nuclei. In addition, when there are different isotopes they will have different scattering lengths. We must therefore allow S to be different for different nuclei even in equivalent positions in the lattice.

For a perfect lattice we obtain in place of (3.12)

$$\frac{e^{ikR}}{R} \sum_j \sum_\mathbf{n} S_{j\mathbf{n}} e^{i\mathbf{q}.(\mathbf{a}_\mathbf{n}+\mathbf{d}_j)}. \qquad (3.38)$$

Let S_j be the average of $S_{j\mathbf{n}}$ over all $\mathbf{n}$, which means averaging over the isotopes with their abundances as weight factors, and over all orientations of the nucleus in each case. Then we may write

$$S_{j\mathbf{n}} = S_j + D_{j\mathbf{n}}, \qquad (3.39)$$

where $D_{j\mathbf{n}}$ vanishes on the average over $\mathbf{n}$. $D_{j\mathbf{n}}$ vanishes also if the nucleus has only one isotope and its spin is zero.

The contribution of S_j to (3.38) is proportional to

$$\sum_j S_j e^{i\mathbf{q}.\mathbf{d}_j} \sum_\mathbf{n} e^{i\mathbf{q}.\mathbf{a}_\mathbf{n}} \qquad (3.40)$$

in close analogy with (3.12). Again the second factor vanishes unless $\mathbf{q}$ equals one of the vectors in the reciprocal lattice (Bragg condition) and the first factor may be used to determine the position of the nuclei in the unit cell, if the strengths of the different diffraction maxima are known. Since S_j is in practice independent of $\mathbf{q}$ (this is because the dimensions of the nucleus are small compared to the wavelength) (3.40) does not tend to become small for high order. Another characteristic feature of the neutron case is that S is appreciable for hydrogen, so that the strength of the reflections depends on the position of the hydrogen atoms or ions, which are hard to locate by X-ray diffraction.

Now consider the contribution from $D_{j\mathbf{n}}$ to (3.38). This vanishes exactly when the Bragg condition is satisfied, since then the exponential factor in (3.38) is the same for all cells, and we are therefore concerned with the average for all N. Hence the D term does not contribute to the diffraction peaks, but only to the continuous background. Since the distribution of the isotopes and of spin orientations over the lattice sites is random, we are interested in the statistical average of the intensity. This is proportional to

$$\sum_{j,j'} \sum_{\mathbf{n},\mathbf{n}'} (S_j + D_{j\mathbf{n}})(S_{j'}^* + D_{j'\mathbf{n}'}^*) e^{i\mathbf{q}\cdot(\mathbf{a_n} - \mathbf{a_{n'}} + \mathbf{d}_j - \mathbf{d}_{j'})}. \tag{3.41}$$

Here the term arising from the product of the S terms is just the square of the Bragg diffraction amplitude (3.40). The terms involving a product of S and D vanish on the average. We are therefore left with the product of the D terms. If in this $\mathbf{n} \neq \mathbf{n}'$, we may still replace each D by its statistical average, since the other factor depends only on the relative position of the cells $\mathbf{n}$ and $\mathbf{n}'$, and averaging over all pairs of cells in the same relative position, we find the distribution of isotopes and spin orientations distributed at random. However, for $\mathbf{n} = \mathbf{n}'$ and $j = j'$ we retain a finite contribution, which equals

$$\sum_{j,\mathbf{n}} D_{j\mathbf{n}}^2 = rN\overline{|D|^2}, \tag{3.42}$$

where, as usual, r is the number of atoms in the unit cell and N the number of cells in the crystal. This quantity is independent of $\mathbf{q}$ and therefore represents a uniform background.

The effect of the thermal motions of the atoms can again be discussed in the same way as in the case of light. For simplicity we shall now assume all nuclei of each element identical and without spin. Then we obtain by an immediate generalization of (3.31) or (3.34) the reduction in intensity of the diffraction maxima, the factor being, for each $\mathbf{q}$, the same as in the case of X-rays.

Similarly, the thermal background is again given by (3.25) with F_j replaced by S_j. However, whereas the energy of an X-ray of wavelength

§ 3.5 NON-CONDUCTING CRYSTALS 73

10^{-8} cm. is about 10 keV, that of a neutron with the same wave vector is about 0·1 eV. While for X-rays the change of energy due to the emission or absorption of one phonon is completely negligible, this is by no means true for neutrons. Consider the case in which the initial neutron wave vector is **k**, and in which a phonon in the state **f**, s is absorbed. Then from (3.26)

$$\mathbf{k'} = \mathbf{k} + \mathbf{f} + \mathbf{K} \tag{3.43}$$

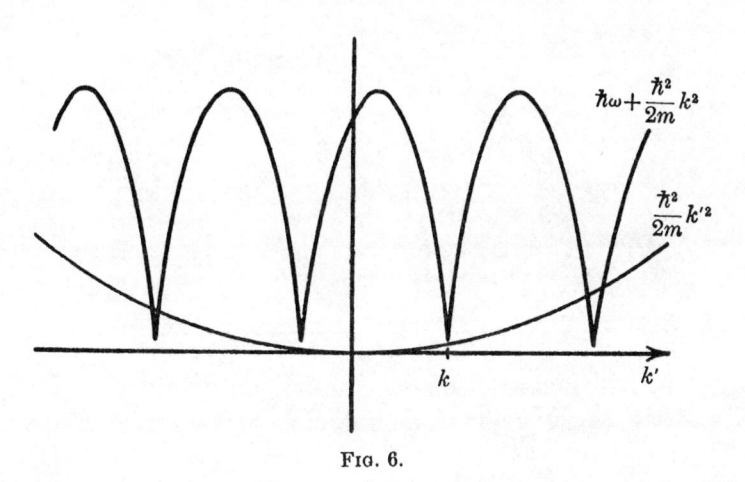

Fig. 6.

and, from energy conservation,

$$\frac{\hbar^2}{2m} k'^2 = \frac{\hbar^2}{2m} k^2 + \hbar\omega(\mathbf{f}, s). \tag{3.44}$$

Fig. 6 shows a graph of the simultaneous equations (3.43) and (3.44) for a one-dimensional case. Each intersection of the parabola with the wavy curve gives a solution. We notice that for the first few intersections we are near the bottom of the vibration spectrum, and f is small, but the increase in neutron energy is appreciable.

Hence the thermal scattering in these circumstances is, for the first few values of K, concentrated in the neighbourhood of the values $k' = k+K$. It is easy to show that this is still true in three dimensions. These fairly sharp peaks must not be confused with diffraction peaks, since for the latter $k'^2 = k^2$ replaces (3.44). For a given orientation of the crystal and given neutron energy there is in general no diffraction from the perfect lattice, but the thermal peaks will be there.

A particularly simple case is that in which k is negligible, i.e. in which we are dealing with incident neutrons of very low velocity. Then the maxima are near $\mathbf{k}' = \mathbf{K}$, i.e. they are always in simple crystallographic directions.

The opposite process of emission of a phonon is similar, and can be discussed by turning the wavy curve of Fig. 6 upside down. For low neutron energy, as in the diagram, there are only few intersections, since the condition that the neutron must lose energy then limits $k'-k$ to be less than $2k$.

IV
ELECTRONS IN A PERFECT LATTICE

4.1. Bloch theorem

IN this and the next few chapters we shall be concerned with the properties of metals. To define a metal, the physicist would refer in the first place to its high electric conductivity, the chemist to its electropositive nature, and the engineer to its ductility and other characteristic mechanical properties. We shall fix our attention on the electric conductivity, and we shall see later that we can understand why substances can be divided into two distinct classes, metals and non-metals, and that the first class has all the characteristic properties that one associates with metals in practice. One can also understand the existence of borderline cases, ranging from 'bad' metals to semi-conductors.

It is well known that the electric conductivity is, in fact, due to the presence of electrons with high mobility, and we are therefore concerned with the problem of an electron moving through a crystal lattice. In this chapter we shall discuss this problem in the case of a perfect lattice, i.e. we shall neglect the lattice vibrations as well as any lattice defects or impurities, and also the effect of any other conduction electrons, except that the space charge due to the other electrons must be allowed for in some average, since we would otherwise find very large and unrealistic electrostatic forces.

Consider, then, the problem of an electron in a periodic field of force, with a potential energy $V(\mathbf{r})$, which must have the translational symmetry of the lattice

$$V(\mathbf{r}+\mathbf{a_n}) = V(\mathbf{r}), \tag{4.1}$$

where $\mathbf{a_n}$ is any of the lattice vectors defined in Chapter I. The Schrödinger equation of the electron is

$$\frac{\hbar^2}{2m}\nabla^2\psi(\mathbf{r})+\{E-V(\mathbf{r})\}\psi(\mathbf{r}) = 0 \tag{4.2}$$

We again choose a cyclic boundary condition to avoid the complications arising from the external surfaces of the crystal,

$$\psi(x+L_1,y,z) = \psi(x,y+L_2,z) = \psi(x,y,z+L_3) = \psi(x,y,z). \tag{4.3}$$

The translational symmetry allows us to classify the wave function in a similar manner to the classification of normal modes in § 1.5. This result, which was derived by Bloch (1928) amongst others, can be obtained in the following manner:

Change the origin of the coordinates in (4.2) by $\mathbf{a_n}$ and use (4.1). It is then evident that $\psi(\mathbf{r}+\mathbf{a_n})$ is also a solution to (4.2). It still satisfies the cyclic condition (4.3). We then have two possibilities: either the energy value E is non-degenerate, so that all solutions of (4.2) must be multiples of each other, or there is degeneracy. In the first case, the new solution is obviously a multiple of the original one,

$$\psi(\mathbf{r}+\mathbf{a_n}) = c\psi(\mathbf{r}). \tag{4.4}$$

The integral of the square modulus of the new function over the crystal is evidently the same as for $\psi(\mathbf{r})$, since it differs merely by a change of origin, and therefore

$$|c|^2 = 1. \tag{4.5}$$

In the case of degeneracy, the general solution can be expressed as an arbitrary linear combination of a basic set of solutions, and the choice of this set is to some degree arbitrary. This freedom is just sufficient to allow us to find a basic set such that each function has the property (4.4). Let the basic set at first be $\psi_1, \psi_2,..., \psi_n$. Then the effect of the displacement on each of them gives again a solution of the equation, which is a linear combination of the basic functions

$$\psi_\mu(\mathbf{r}+\mathbf{a_n}) = \sum_{\nu=1}^{n} C_{\mu\nu}\psi_\nu(\mathbf{r}), \qquad \mu = 1, 2,..., n.$$

We may assume the original set orthogonal and normalized,

$$\int \psi_\mu^*(\mathbf{r})\psi_\nu(\mathbf{r}) \, d^3r = \delta_{\mu\nu},$$

and, since these integrals are not affected by a change of the origin, the same will be true of the new functions. Hence

$$\sum C_{\nu\mu}^* C_{\mu\lambda} = \delta_{\nu\lambda},$$

which shows that the matrix C is unitary. It is well known that any unitary matrix can, by a unitary transformation, be brought to diagonal form. There exists, therefore, a set of linear combinations of the ψ which are again orthogonal and normalized, and which belong to a diagonal matrix C. This is equivalent to (4.4) being valid for each of the functions.

We would like (4.4) to apply for an arbitrary lattice vector $\mathbf{a_n}$ (the constant c will of course depend on $\mathbf{n}$) and we must therefore make sure that, having made C diagonal for a particular displacement $\mathbf{a_n}$, we are still free to achieve the same for other displacements without destroying what we have already achieved. This follows from the fact that any two displacements are commuting operations, i.e. the result is the same regardless of the order in which the operations are carried out. This

§ 4.1 ELECTRONS IN A PERFECT LATTICE

shows that the matrices C belonging to any two lattice vectors must commute, and it is well known that a set of commuting matrices can simultaneously be brought to diagonal form.

We may therefore assume, from now on, that all eigenfunctions are chosen so as to satisfy the condition (4.4) for any displacement.

The displacement $\mathbf{a_n}+\mathbf{a_{n'}}$, which is the result of two successive displacements, multiplies the wave function by cc'. Hence $\log c$ is additive, and must therefore depend linearly on the displacement vectors, which shows, in conjunction with (4.5), that

$$c = e^{i\mathbf{k}.\mathbf{a_n}}, \tag{4.6}$$

so that
$$\psi(\mathbf{r}+\mathbf{a_n}) = e^{i\mathbf{k}.\mathbf{a_n}}\psi(\mathbf{r}). \tag{4.7}$$

The vector $\mathbf{k}$ is defined only apart from an arbitrary vector $\mathbf{K}$ of the reciprocal lattice, since, by the definition of the latter,

$$e^{i\mathbf{K}.\mathbf{a_n}} = 1 \tag{4.8}$$

for any $\mathbf{n}$ (1.28). We may use this freedom again to choose $\mathbf{k}$ within the basic cell of the reciprocal lattice, defined in § 6.1, which in this connexion is often called the basic 'Brillouin zone'.

The cyclic condition again restricts the components of $\mathbf{k}$ to be multiples of $2\pi/L_1$, $2\pi/L_2$, and $2\pi/L_3$, respectively. The permitted vectors $\mathbf{k}$ are thus the same as the wave vectors $\mathbf{f}$ of Chapter I, and their number is equal to the number of unit cells in the crystal.

We may therefore label the wave functions by the corresponding vector $\mathbf{k}$. There will, of course, be an infinite series of wave functions belonging to the same $\mathbf{k}$, and we shall call these $\psi_{kl}(\mathbf{r})$ and the corresponding energy eigenvalues $E_l(\mathbf{k})$. The eigenvalues belonging to the same $\mathbf{k}$ are all different, except in very special cases.

The property (4.7) may also, following Bloch, be expressed in a different form. Let

$$\psi_{kl}(\mathbf{r}) = e^{i\mathbf{k}.\mathbf{r}}u_{kl}(\mathbf{r}), \tag{4.9}$$

where u is defined by this equation. Then (4.7) shows that

$$u_{kl}(\mathbf{r}+\mathbf{a_n}) = u_{kl}(\mathbf{r}). \tag{4.10}$$

In other words, $u_{kl}(\mathbf{r})$ has the translational symmetry of the lattice.

This function satisfies the equation

$$\left[\frac{\hbar^2}{2m}(\nabla-i\mathbf{k})^2+\{E_l(\mathbf{k})-V(\mathbf{r})\}\right]u_{kl}(\mathbf{r}) = 0, \tag{4.11}$$

which, because of the periodicity of $u_{kl}(\mathbf{r})$, need only be solved for one unit cell, with boundary conditions on opposite faces of the cell, to make the periodic continuation possible.

We see at once that by going to complex conjugate quantities we have the same equation for $-\mathbf{k}$ with u^* as solution. Hence

$$u_{-\mathbf{k}l}(\mathbf{r}) = u_{\mathbf{k}l}^*(\mathbf{r}); \qquad E_l(-\mathbf{k}) = E_l(\mathbf{k}). \tag{4.12}$$

One also knows that a small change in the coefficients of a differential equation will cause a small change in its solution. If we assign the label correctly, $E_l(\mathbf{k})$ must therefore be a continuous function of $\mathbf{k}$, and have a continuous derivative. (It can, in fact, be shown to be an analytic function of $\mathbf{k}$, but we do not need to use this.)

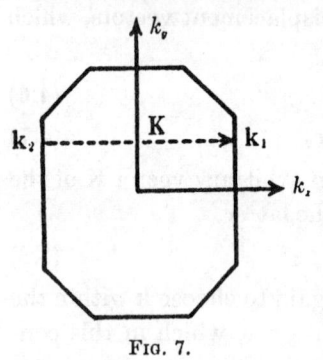

This continuity also applies as between points on opposite faces of the basic cell in $\mathbf{k}$ space, if we remember that the location of $\mathbf{k}$ values in this cell was purely a matter of convention. This is illustrated by Fig. 7 for the plane case. The polygon indicates the basic cell, the point $\mathbf{k}_1$ a value on the boundary. Now this differs from $\mathbf{k}_2$ on the opposite boundary just by a reciprocal lattice vector, $\mathbf{K}$. Points to the immediate right of $\mathbf{k}_2$ might just as well have been placed to the immediate right of $\mathbf{k}_1$. This shows that

Fig. 7.

$$E_l(\mathbf{k}_1) = E_l(\mathbf{k}_2); \quad \operatorname{grad} E_l(\mathbf{k}_1) = \operatorname{grad} E_l(\mathbf{k}_2). \tag{4.13}$$

Now suppose that the symmetry of the lattice permits a reflection in the plane $x = 0$. From this we may conclude that

$$E_l(\mathbf{k}_1) = E_l(\mathbf{k}_2); \qquad \frac{\partial E_l(\mathbf{k}_1)}{\partial k_y} = \frac{\partial E_l(\mathbf{k}_2)}{\partial k_y};$$

$$\frac{\partial E_l(\mathbf{k}_1)}{\partial k_x} = -\frac{\partial E_l(\mathbf{k}_2)}{\partial k_x}. \tag{4.14}$$

The first two statements are identical with those derived before from continuity, but the last one differs in sign and hence

$$\frac{\partial E_l(\mathbf{k}_1)}{\partial k_x} = 0. \tag{4.15}$$

In general, provided the lattice has sufficient symmetry, the derivative of E in a direction normal to a cell boundary vanishes on that boundary. This is clearly also true in three dimensions.

There is a possible exception to this rule when two different functions, say $E_1(\mathbf{k})$ and $E_2(\mathbf{k})$, happen to be equal on the boundary. In that case

(4.13) need not apply; it may instead be that the gradient of E_1 at $\mathbf{k}_1$ equals the gradient of the other function E_2 at $\mathbf{k}_2$. We shall meet an example of this later, but shall also see that it is very exceptional.

Fig. 8 shows, for the case of a linear chain, some typical $E(k)$ curves which are compatible with the results so far obtained.†

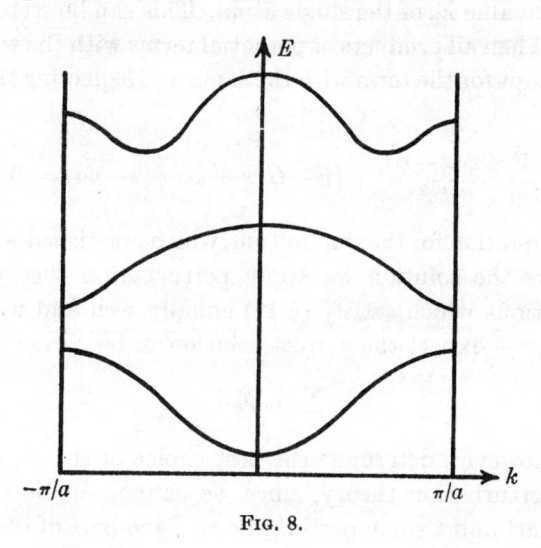

Fig. 8.

4.2. Strong binding

So far we have made no assumption about the potential function $V(\mathbf{r})$ beyond its periodicity, but one may gain further insight into the problem by solving the equations for two limiting cases, that of strongly bound and that of nearly free electrons.

In the limit of strong binding, we assume the lattice constant so large, or the atomic radius so small, that the wave functions of atoms at adjacent lattice points would overlap very little.

To avoid inessential complications, we assume a linear problem, and identical atoms. We may then write the potential in the form

$$V(x) = \sum_n U(x-na), \qquad (4.16)$$

where $U(x)$ is the potential energy of the electron near an atom at the origin, and the sum goes over all atoms in the chain. We then expect that the eigenfunction will bear some relation to the eigenfunction $\phi(x)$

† Dr. Shockley has pointed out to me that one can prove in the one-dimensional case that the minima or maxima of $E(k)$ may occur only at $k = 0$ or $k = \pm\pi/a$. The upper curve of Fig. 8 cannot therefore be realized in one dimension, but it may occur for a linear chain of three-dimensional atoms.

of a single atom. Indeed, any of the N functions $\phi_n(x) = \phi(x-na)$ is very nearly a solution of the wave equation

$$\frac{\hbar^2}{2m}\frac{\partial^2 \psi(x)}{\partial x^2} + \left\{E - \sum_n U(x-na)\right\}\psi(x) = 0 \qquad (4.17)$$

with the eigenvalue E_0 of the single atom. This can be seen by inserting ϕ_n in (4.17). Then all products of potential terms with the wave function are small, except for the term with the same n. Neglecting the others, we are left with

$$\frac{\hbar^2}{2m}\frac{d^2\phi(x-na)}{dx^2} + \{E - U(x-na)\}\phi(x-na) = 0, \qquad (4.18)$$

which is the equation for the single atom, which is satisfied when $E = E_0$.

To improve the solution we apply perturbation theory. Since we have N functions which satisfy (4.17) equally well and with the same energy value, we expect the correct solution to be close to some linear combination

$$\sum_n A_n \phi_n. \qquad (4.19)$$

We cannot, however, determine the best choice of the A_n by the usual method of perturbation theory, since we cannot divide the potential into a large part and a small perturbation. Each part of the potential is large for one of the wave functions contained in (4.19).

Instead we select the coefficients of the linear combination (4.19) by using the variation principle; we choose the coefficients A_n in such a way as to make the expectation value of the energy stationary. This expectation value is

$$E = \frac{\sum_{n,m} A_n^* A_m H_{n,m}}{\sum_{n,m} A_n^* A_m J_{n,m}}, \qquad (4.20)$$

where

$$J_{n,m} = \int \phi_n \phi_m \, dx; \qquad H_{n,m} = \int \phi_n \left(-\frac{\hbar^2}{2m}\frac{d^2}{dx^2} + V\right)\phi_m \, dx. \qquad (4.21)$$

We note, first of all, that $J_{n,m}$ and $H_{n,m}$ depend only on the difference of the suffixes $n-m$, as one can see at once by a change of origin in the integrals (4.21), and that $J_{n,m} = J_{m,n}$, $H_{m,n} = H_{n,m}$. We may therefore denote them by $J(n-m)$ and $H(n-m)$.

Using (4.18), one finds that

$$H(n-m) = E_0 J(n-m) + \int \phi_n \{V(x) - U(x-ma)\}\phi_m \, dx$$

$$= E_0 J(n-m) + h(n-m). \qquad (4.22)$$

§ 4.2   ELECTRONS IN A PERFECT LATTICE

$h(n-m)$ contains all potentials except that of the mth atom, and since the integrand contains ϕ_m as a factor, the integral is small. $J(n-m)$ is small, unless $n = m$.

The condition that (4.20) be stationary is

$$\sum_m \{H(n-m) - EJ(n-m)\} A_m = 0,$$

or, using (4.22),

$$\sum_m \{h(n-m) - (E-E_0)J(n-m)\} A_m = 0. \qquad (4.23)$$

We can now find a solution in the form

$$A_n = A e^{i\kappa n}. \qquad (4.24)$$

Insertion in (4.23) shows that this is satisfied, provided

$$E - E_0 = \frac{\sum_n h(n) e^{i\kappa n}}{\sum_n J(n) e^{i\kappa n}}. \qquad (4.25)$$

Now both $h(n)$ and $J(n)$ decrease very rapidly with increasing n, since they contain the product of two atomic wave functions displaced a distance na relative to each other. However, $h(0)$ and $h(1)$ are of comparable magnitude and $J(0)$ is unity if ϕ is normalized. Then, to the leading order,

$$E - E_0 = h(0) + 2h(1)\cos\kappa. \qquad (4.26)$$

Our 'zero-order' wave function (4.19) with the coefficients (4.24) evidently satisfies (4.7) with

$$\kappa = ka. \qquad (4.27)$$

The first-order energy (4.26) is similar to the lower two curves sketched in Fig. 8.

We have not at any point assumed that the eigenvalue E_0 represents the lowest energy of the atom, and the procedure can be carried out for each atomic level. In this way we obtain a set of energy levels $E_l(k)$ and corresponding wave functions, which form a complete orthogonal set. They are therefore suitable for improving the approximation further by perturbation theory of higher order if required.

In this approximation each of the atomic levels is drawn out into a continuous band of width $4h(1)$. This width must be small compared to the spacing of the atomic levels if our approximations are justifiable.

We have here tacitly assumed that the atomic state is non-degenerate, which, in fact, is usually the case in one dimension. If we are dealing with a degenerate level of s components, our wave function will have to be taken as a combination of sN different atomic functions. After splitting off an exponential factor (4.24) we are still left with a set of s-fold simultaneous equations. I shall not cover this case in detail.

These results are immediately generalized to three dimensions. The zero-order wave function takes the form

$$\psi_{k,l}(r) = \sum_{n,j} A_j e^{i\mathbf{k}\cdot\mathbf{a}_n} \Phi_l(\mathbf{r}-\mathbf{a}_n-\mathbf{d}_j), \qquad (4.28)$$

where $\Phi_l(\mathbf{r})$ is again a single-centre eigenfunction. In a simple lattice, with one atom per unit cell, there is no summation over j, and (4.28) determines the solution completely. In an element with a more complex lattice, A_j has still to be obtained from a set of r simultaneous equations. If the atomic state is degenerate, the number of equations is further increased. For a crystal containing different species of atoms, the sum over j should cover only the atoms of one kind, since the unperturbed energy levels of different atoms are in general different, and the corresponding states not degenerate.

In the simplest case of $r = 1$, and an atomic s-level, (4.25) becomes

$$E - E_0 = \frac{\sum_n h(\mathbf{n}) e^{i\mathbf{k}\cdot\mathbf{a}_n}}{\sum_n J(\mathbf{n}) e^{i\mathbf{k}\cdot\mathbf{a}_n}}, \qquad (4.29)$$

where the meaning of the coefficients is evident by a simple generalization of (4.21) and (4.22).

For example, for a body-centred cubic lattice, for which the reciprocal lattice is face-centred cubic, and has as basic cell a rhombohedron, the leading terms in (4.29), corresponding to the approximation (4.26), give

$$E - E_0 = h_0 + 8h_1 \cos \tfrac{1}{2}k_x a \cos \tfrac{1}{2}k_y a \cos \tfrac{1}{2}k_z a. \qquad (4.30)$$

Fig. 9 shows two sections through the basic reciprocal cell, for $k_z = 0$, and $k_z = \pi/2a$, respectively, with lines of constant energy to the approximation (4.30). Near $\mathbf{k} = 0$, the surfaces of constant energy are spheres and, for negative h_1, this region corresponds to the energy minimum. The highest energy is attained at the corners, for which any one of the components of $\mathbf{k}$ is $2\pi/a$, and the others zero. These points are all equivalent. In their neighbourhood the surfaces of constant energy are parts of spheres, which may be thought of as forming a continuous sphere, cut into octants by our convention about the choice of $\mathbf{k}$.

Between these extremes, the energy surfaces are more complicated in shape. For the special case $E = E_0 + h_0$, the surface is a cube.

From a physical point of view, the approximation of this section may be pictured in the following way: For tight binding an electron is almost in a stationary state when it describes an orbit near one of the attractive centres. There is, however, a probability of a transition in which it changes to an orbit about the adjacent centre, in spite of the fact that there is a potential barrier between them (tunnel effect). It follows that the electron will progress slowly through the lattice. What is less obvious from this picture is the fact that in a perfect lattice this motion has associated with it the wave vector **k**, which is a constant of the motion and which, as we shall see, results in the electron progressing, on the average, with uniform velocity.

In a real crystal, this approximation would be reasonable for the electrons in the inner shells, whose radius is small compared to the inter-atomic distances, so that the integrals h and J are small. These levels will be drawn out into bands whose width is very small compared to the distances between them. In the outer shells, which are of interest for the conduction electrons, the overlap will be large, and the results of this section have at best qualitative significance.

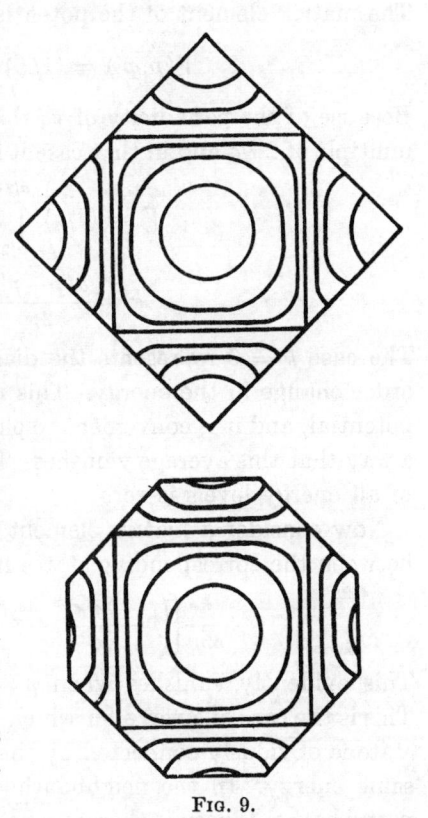

Fig. 9.

4.3. Nearly free electrons

One may also solve the problem in the opposite limit of nearly free electrons. We assume the potential $V(\mathbf{r})$ to be weak, so that we may start from free electrons and then apply a perturbation.

The condition for the validity of perturbation theory is that the matrix elements of the perturbing potential be small compared to the energy differences between the levels to which these matrix elements refer. We again begin, for simplicity, with the one-dimensional case.

The unperturbed wave functions are

$$\psi_p = \frac{1}{\sqrt{L}} e^{ipx}, \tag{4.31}$$

where L is the length of the chain, and p the momentum measured in units of $\hbar$. The energy is

$$E(p) = \frac{\hbar^2}{2m}p^2. \tag{4.32}$$

The matrix element of the potential between two states p, p' is

$$V(p,p') = (1/L) \int V(x) e^{i(p-p')x}\, dx.$$

Because of the periodicity of V, this integral vanishes unless $p'-p$ is a multiple of $2\pi/a$ and in that case it is

$$V_\kappa = (1/a) \int_{-a/2}^{a/2} V(x) e^{2\pi i \kappa x/a}\, dx,$$

$$\kappa = \frac{a(p-p')}{2\pi} = 0,\ \pm 1,\ \pm 2,\ldots. \tag{4.33}$$

The case $\kappa = 0$ represents the diagonal element, and hence the first-order change in the energy. This is equal to the space average of the potential, and it is convenient to choose the zero level of energy in such a way that this average vanishes. Then $V_0 = 0$, and the first-order shift of all energy levels is zero.

Now consider a matrix element with $\kappa \neq 0$. The energy difference between the corresponding states is

$$\frac{\hbar^2}{2m}\left\{\left(p-\frac{2\pi\kappa}{a}\right)^2 - p^2\right\} = \frac{2\pi\hbar^2\kappa}{ma}\left(\frac{\pi\kappa}{a}-p\right). \tag{4.34}$$

This evidently vanishes when $p = \pi\kappa/a$, and in that case $p' = -p$. This is the case of a wave for which Bragg scattering is possible, and this state is obviously connected by the potential with another state of the same energy. In the neighbourhood of such states the condition for perturbation theory is therefore never satisfied. On the other hand, for states well away from these exceptional cases, the quantity (4.34) is finite, and we may postulate that it is larger than V_κ. In that region, the energy shift is of second order in V, precisely

$$\Delta E = \sum_\kappa \frac{|V_\kappa|^2}{E(p) - E\{p - (2\pi\kappa/a)\}} \tag{4.35}$$

apart from terms of higher order.

In the neighbourhood of the Bragg reflection of order κ, we may assume that all but one of the energy differences appearing in (4.35) are large. We may then apply the perturbation theory for nearly degenerate states by assuming that the energy difference between the two states in question is comparable with the matrix element V. Denoting the

§4.3 ELECTRONS IN A PERFECT LATTICE

two states for brevity by 1 and 2, and their unperturbed energies by E_1 and E_2, we then try to find a solution of the perturbed equation which lies close to
$$A_1\psi_1 + A_2\psi_2. \tag{4.36}$$
The coefficients A_1, A_2 are to be determined in such a manner as to diagonalize that part of the total energy operator whose matrix elements

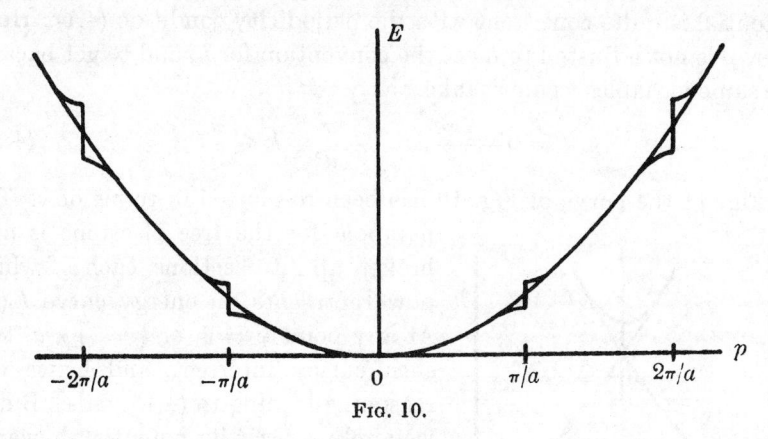

Fig. 10.

do not connect the two chosen states with any others. This leads to the equations
$$\left. \begin{array}{l} (E-E_1)A_1 + V_\kappa A_2 = 0 \\ V_\kappa^* A_1 + (E-E_2)A_2 = 0 \end{array} \right\} \tag{4.37}$$
which have a solution if the determinant vanishes. This requires
$$E = \tfrac{1}{2}(E_1+E_2) \pm \sqrt{\left\{\left(\frac{E_1-E_2}{2}\right)^2 + |V_\kappa|^2\right\}}. \tag{4.38}$$

If the energy difference is large compared with V_κ, then the two energy values resulting from this lie close to E_1 and E_2 respectively, except for terms of the order $|V_\kappa|^2/(E_1-E_2)$, which may then be taken together with the terms from transitions to other states. If, on the other hand, $E_1 = E_2$, as for exact Bragg reflection, then the two values of (4.38) differ by $2|V_\kappa|$. For small difference $E_1 - E_2$ the gap between the perturbed energy values differs from $2|V_\kappa|$ only by an amount proportional to the square of $E_1 - E_2$.

Fig. 10 indicates how the energy of free electrons becomes modified by a small disturbance of this kind. The cases $\kappa = \pm 1, \pm 2$ are shown.

It is easy to see from (4.37) that for $E_1 = E_2$, i.e. for points on either side of the discontinuity, $|A_1/A_2| = 1$. The wave functions (4.36) are therefore combinations of two plane waves (4.31) travelling in opposite directions with equal amplitude, and thus are standing waves. The one with the lower energy has its maximum amplitude at the centres, where the potential is most attractive, and the one with the higher energy has its nodes there.

We must now connect these results with the general theory of § 4.1. We note, for this purpose, that already the unperturbed wave (4.31) satisfies the theorem (4.7), which is not surprising, since the absence of a potential is quite consistent with the periodicity condition (4.1). However, p is not adjusted to meet the convention for k, and to get back to the same variables we must take k as

$$k = p - \frac{2\pi\kappa}{a}, \qquad -\frac{\pi}{a} \leqslant k \leqslant \frac{\pi}{a}. \qquad (4.39)$$

In Fig. 11 the curve of Fig. 10 has been re-plotted in terms of k. The parabola for the free electrons is now broken up into sections, each of which now represents an energy curve $E_l(k)$. At any point $k = 0$, or $k = \pm\pi/a$, two such curves intersect, and hence the reasoning leading to (4.15) fails. But if we apply a periodic potential, however weak, the intersection disappears, and the curve has zero slopes at the zone boundary.

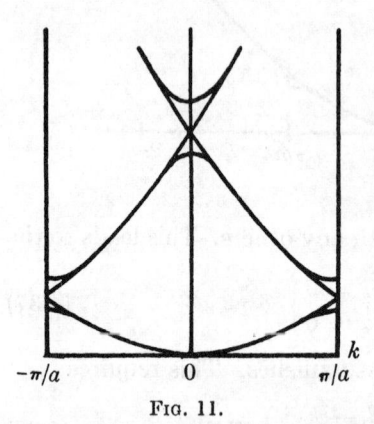

Fig. 11.

We see therefore that the qualitative appearance of the energy curves is very similar in the two extreme limits of very strong and very weak binding, the principal difference being that in the one case we have narrow bands widely separated from each other, and in the other wide energy bands separated by small gaps. Another important difference is the uneven curvature of the curves of Fig. 11, with its very narrow turns near the boundaries.

It is evident that the present method is best at high energies, since we have neglected the matrix elements of V in comparison with the vertical distance of the various curves of Fig. 11, except the nearest one. These vertical distances increase as we go upwards, whereas the matrix elements (4.33) decrease with increasing order.

Our present results also give a convenient method for discussing electron diffraction. If an external beam of electrons is incident on a crystal, and its energy lies in one of the energy gaps appearing in Fig. 10 or 11, it must be totally reflected, since there is no possible motion of an electron inside the crystal with that energy. By the method of Chapter

III we should have calculated an infinitely intense scattered wave for a sharply defined wavelength, and hence for a single value of the energy. Instead, we now find a finite reflected amplitude (equal to the amplitude of the wave incident on the crystal) but over a finite range of energies. This result is equivalent to Ewald's dynamical theory of diffraction, which was developed for the X-ray case. For a quantitative treatment, surface effects must, however, also be taken into consideration.

The generalization of the treatment of nearly free electrons to three dimensions presents little difficulty. In momentum space the condition

$$(\mathbf{p}-\mathbf{K})^2 = \mathbf{p}^2, \qquad (4.40)$$

which is just a form of the Bragg condition, represents, for each reciprocal lattice vector $\mathbf{K}$, a plane across which a discontinuity will appear. These planes dissect momentum space into polyhedra, of which the one containing the origin is the 'basic Brillouin zone'. Of the others we can always find a set which, after displacements by suitable reciprocal vectors $\mathbf{K}$, will together just cover the basic zone again, and then belong to a continuous energy function. The polyhedra, which together make up such a set, are known as Brillouin zones.

4.4. Velocity and acceleration

Since the wave function $\psi_l(\mathbf{k})$ of a particular electron state is complex, we would in general expect it to have a non-vanishing mean velocity. This can be derived easily from the equation of continuity. Suppose we are not dealing with a strictly stationary state, but with a wave packet made up of a number of waves with adjacent values of $\mathbf{k}$. If the spread of $\mathbf{k}$ values used in building the wave packet is to be small compared to the dimensions of the basic Brillouin zone, the extent of the wave packet in space must be large compared to the size of a lattice cell, but it may still be very small compared to the size of the crystal.

We can then find the displacement in time of the mean position of the packet. It must, by a well-known argument, equal the group velocity

$$\mathbf{v} = \frac{\partial \omega}{\partial \mathbf{k}} = \frac{1}{\hbar} \frac{\partial E_l(\mathbf{k})}{\partial \mathbf{k}}. \qquad (4.41)$$

(By the derivative with respect to the vector $\mathbf{k}$ we mean that each component of $\mathbf{v}$ equals the derivative with respect to the corresponding component of $\mathbf{k}$.) Now the rate at which the wave packet travels must equal the transport in the wave packet, and, since the states from which it is built are very similar, we may identify the mean transport with the transport for each of them. Hence (4.41) gives the transport, or the average velocity, for a state **k**. From (4.15) it follows that the normal velocity component vanishes at a zone boundary, in line with the fact that at such a boundary Bragg reflection makes any progress of the electron impossible.

Next consider the action of an applied electric field. This will induce transitions from one stationary state to others. In the weak fields that, in practice, can be applied to metals, we may disregard transitions in which l changes, since this will involve a much greater change of the electron energy than the field can supply with any reasonable probability (this is not the case when one is concerned with the electric breakdown of insulators). We may therefore suppose that the only effect of the field is to alter the value of **k**. Now if **k** is the mean wave vector of a wave packet, the change in the electron energy with time is

$$\frac{\partial E_l}{\partial \mathbf{k}} \cdot \frac{d\mathbf{k}}{dt}.$$

On the other hand, the work done by the field per unit time is

$$e\mathbf{v}\cdot\mathbf{F} = \frac{e}{\hbar}\frac{\partial E_l(\mathbf{k})}{\partial \mathbf{k}}\cdot\mathbf{F},$$

where the second expression is taken from (4.41). From energy conservation these two quantities must be equal, and hence

$$\frac{d\mathbf{k}}{dt} = \frac{e}{\hbar}\mathbf{F} \tag{4.42}$$

must give the rate of change of **k** in the field. This result is obvious for free electrons, for which $\hbar\mathbf{k}$ is the momentum, apart from an additive constant.

That it should always hold is at first sight surprising if you look at curves like those of Figs. 8 or 11, which show that there are always regions in which the slope of the energy curve decreases with increasing **k**. Since this slope measures the mean velocity, it follows that it is possible for the field to accelerate the electron in a direction opposite to the usual. This is less strange if we remember the connexion with diffraction. What happens is that the field pushes the electron near to the state in which complete reflection brings it, in the mean, to rest.

By analogy one would expect that the effect of a magnetic field would be to change **k** at the rate

$$\frac{d\mathbf{k}}{dt} = \frac{e}{\hbar c}\mathbf{v}\wedge\mathbf{H}, \tag{4.43}$$

where c is the velocity of light, **v** the velocity (4.41) and the sign $\wedge$ indicates a vector product. This formula is indeed correct for many purposes. It neglects, however, the fact that, in a magnetic field, the electron describes closed orbits (in the plane at right angles to the field direction) and that therefore the energy levels will be in part discrete. This fact, which is important for the theory of diamagnetism, cannot be expressed in terms of an acceleration; we shall return to this problem at a later stage.

4.5. Many electrons. Statistics

We are now ready to consider the distribution of electrons over the stationary states we have previously described. To do this rigorously one would have to allow for the interaction of the electrons with each other, but this would give rise to formidable mathematical difficulties. The inability to include the electron interaction is the greatest source of uncertainty in the application of the electron theory of metals. In certain problems we shall be able to see what the effect of the electron interaction would be and even to allow for it approximately, but in many cases a large measure of doubt must remain.

For the present we assume, therefore, that the electrons do not interact, except that their mean charge distribution is taken into account in the definition of our potential.

We must, however, take into account Pauli's exclusion principle, according to which any one quantum state can contain at most two electrons of opposite spin. The state of lowest energy of the whole system is therefore obtained by filling the lowest $Nrz/2$ states, where, as before, N is the number of unit cells in the crystal, r the number of atoms in a cell, and z the number of electrons per atom. The energy of the highest state so occupied will be called η_0. Now there are two distinct possibilities. Either η_0 coincides with the upper end of an energy band, so that this band is completely filled and the next higher one empty, or η_0 lies inside a band.

In the first case, the state of lowest energy is not the beginning of a continuous energy spectrum for the whole crystal, but is isolated from all high states by a finite gap. This is evident since no other arrangement of the electrons is possible without moving at least one of them up into the next band, which takes a finite amount of energy. Such a substance could not respond to any weak external disturbance, and could not, for example, redistribute the electrons to cancel an external potential difference. It must therefore be an insulator. On the other hand, if there are free states in the last band, it is possible to shift the electrons to states of slightly different energy so as to set up a net current, or, by the formation of wave packets, produce a non-uniform charge distribution. This is the case of a metal.

In the case of one atom per unit cell, where the number of states in a band equals the number of atoms, it is not possible to fill a number of bands completely, unless the number of electrons per atom is even. All elements of odd atomic number with a structure for which $r = 1$ should be metals.

At first sight one would expect the converse to be true, that all even elements with $r = 1$ should be non-metals. However, this is not correct, since the bands may not be distinct in energy. It is true that for given $\mathbf{k}$ the different energy values $E_l(\mathbf{k})$ are in general distinct, but the highest value of $E_l(\mathbf{k})$ may well be above the minimum of E_{l+1}. This will always be the case at very high energy, where from the work of § 4.3 we saw that the energy gaps at a zone boundary are very small, whereas the variation of energy along the boundary will be appreciable.

Amongst elements with simple structures the metals are therefore predominant, but in the case of more complex structures the number of states in a band is only $1/r$ times the number of atoms, and it is then more likely that they will be separated in energy. We shall see later that there is, in fact, a tendency to favour a structure for which this happens to be the case.

Bands which are completely filled may in general be left out of consideration completely in the discussion of conduction phenomena and the like. These certainly include the states corresponding to the innermost atomic shells. We shall refer to the incompletely filled band as the conduction band. From the above discussion it is clear that there may be, and in a metal of even atomic number with $r = 1$ there must be, more than one conduction band.

A particularly simple case is that in which a certain band contains a small number of electrons. These will then move in states near the bottom of the band, where the energy may be regarded as a quadratic function of $\mathbf{k}$. If this energy minimum is attained for only a single value of $\mathbf{k}$, then the energy surface in its neighbourhood must have the same symmetry as the structure. If the crystal is cubic, the quadratic function must be isotropic and therefore

$$E(\mathbf{k}) = \text{constant} + b(\mathbf{k} - \mathbf{k}_0)^2 \qquad (4.44)$$

with constant b. Apart from the change in origin, this is similar to the energy function for free electrons, and we may write

$$b = \frac{\hbar^2}{2m^*},$$

where m^* is of the dimension of a mass, and may be called the effective mass. In such states, the behaviour of electrons is in many respects the same as that of free charged particles of mass m^*. It may also be, however, that the energy function, which has the crystal symmetry, reaches its lowest value at a number of points which are symmetrically placed. In that case the function is still quadratic near each of those points, but we can no longer claim that it must be isotropic. The surfaces of constant energy near one minimum may then be ellipsoids, and it merely follows, for a cubic crystal, that a rotation by 90° about a crystal axis must convert this ellipsoid into the corresponding ellipsoid at another minimum.

An equally simple case is that in which the band is almost full, with only a small fraction of the states vacant. In that case it is convenient to specify the vacant states rather than the occupied ones. Since the full band has no conductivity, all conduction processes may be described as being due to the presence of this small number of holes from which electrons have been removed, as compared to a completely filled band. The absence of an electron means the addition of a positive charge. An electric field in the x-direction will then shift all electrons to states of increased k_x, according to (4.42), which, by (4.41), near an energy maximum means decreasing velocity. The vacant states or holes are also now moving with decreasing velocity in the x-direction, as if they were real positive charges. The absence of electrons travelling in the $-x$-direction represents, of course, a current in the positive x-direction, so that the current is in the direction of the field. The energy required to produce a vacancy is, of course, a minimum for the state of highest energy, and if, near a single maximum, the energy is

$$\text{constant} - b(\mathbf{k} - \mathbf{k}_0)^2,$$

the constant b can be linked to an effective mass of the positive holes as before. It is hardly necessary to point out the similarity of this reasoning to Dirac's theory of the positron.

At a finite temperature, when the system is not in the state of lowest energy, the electron distribution is governed by Fermi-Dirac statistics. The average number of electrons in a state of energy E is given by

$$2f(E) = \frac{2}{1+e^{(E-\eta)/kT}}, \tag{4.45}$$

and the free energy is

$$F = N_e \eta - 2kT \sum_i \log\{1 + e^{-(E_i - \eta)/kT}\}, \tag{4.46}$$

where N_e is the total number of electrons, and i labels the states, E_i being the corresponding energies. The factor 2 in (4.45) and (4.46) takes account of the two spin directions.

The constant η has to be determined in such a way as to make the total number of electrons in all states equal to N_e, or

$$N_e = \sum_i 2f(E_i). \tag{4.47}$$

In practice, kT is very small compared to the width of a band, and compared to the width of the filled and the vacant parts of the band, unless the number of conduction electrons or vacancies is exceptionally small. We may then apply the usual approximation for a highly degenerate Fermi gas.

In particular, when we are dealing with an integral over the Fermi function of the form

$$\int_{-\infty}^{\infty} dE \frac{dG}{dE} f(E), \tag{4.48}$$

where G is some function of the energy which is smooth near $E = \eta$, we may integrate by parts to obtain

$$-G(-\infty) - \int_{-\infty}^{\infty} dE\, G(E) \frac{df}{dE}.$$

Now df/dE is greatest near $E = \eta$, and vanishes rapidly if $|E-\eta|$ exceeds kT. It is therefore convenient to expand G in a Taylor series near η, and we obtain

$$-G(-\infty) + G(\eta) + G'(\eta)f_1 + \tfrac{1}{2}G''(\eta)f_2 + \tfrac{1}{6}G'''(\eta)f_3 + \ldots,$$

where the accents denote derivatives, and $f_1, f_2, \ldots$ the moments

$$f_n = -\int_{-\infty}^{\infty} dE(E-\eta)^n \frac{df}{dE}.$$

Now since

$$-\frac{df}{dE} = \frac{1}{kT} \frac{1}{(1+e^{(E-\eta)/kT})(1+e^{-(E-\eta)/kT})} \tag{4.49}$$

is evidently an even function of $(E-\eta)$, all odd moments $f_1, f_3, \ldots$ vanish. The second moment f_2 is

$$f_2 = \frac{\pi^2}{3}(kT)^2,$$

so that, apart from terms of order T^4 or higher, the required integral is

$$\int_{-\infty}^{\infty} dE \frac{dG}{dE} f(E) = G(\eta) - G(-\infty) + \frac{\pi^2}{6}(kT)^2 G''(\eta). \tag{4.50}$$

This result may be used to evaluate condition (4.47), which may be written in integral form, if $Z(E)$ denotes the number of energy levels below E,

$$N_e = 2 \int_{-\infty}^{\infty} dE \frac{dZ}{dE} f(E) = 2Z(\eta) + \frac{\pi^2}{3}\left(\frac{d^2Z}{dE^2}\right)_\eta (kT)^2. \qquad (4.51)$$

Hence we can find the temperature dependence of η,

$$\eta - \eta_0 = -\frac{(d^2Z/dE^2)_\eta}{(dZ/dE)_\eta} \frac{\pi^2}{6}(kT)^2. \qquad (4.52)$$

The ratio of the first to the second derivative of Z is the size of the energy interval over which the density of states changes appreciably. As long as this is larger than kT, $\eta - \eta_0$ is small compared to kT, and may therefore for most purposes be neglected.

4.6. Specific heat

The energy content of the electron system is easily calculated from the results of the preceding section.

The total energy is evidently

$$E(T) = \sum_i E_i f(E_i), \qquad (4.53)$$

where i again numbers all possible electron states. Introducing again the number $Z(E)$ of states with energy below E,

$$E(T) = 2 \int_{-\infty}^{\infty} f(E) E \frac{dZ}{dE} dE, \qquad (4.54)$$

and, using (4.50), with

$$G(E) = \int_{-\infty}^{E} E \frac{dZ}{dE} dE, \qquad (4.55)$$

we find $\qquad E(T) = 2 \int_{-\infty}^{\eta} E \frac{dZ}{dE} dE + \frac{\pi^2}{3}(kT)^2 \left(\frac{dZ}{dE} + E \frac{d^2Z}{dE^2}\right)_\eta.$

We have already seen that η is very little different from η_0, and therefore we may take the first integral to the limit η_0 instead of η and correct to first order in the difference:

$$E(T) = 2 \int_{-\infty}^{\eta_0} E \frac{dZ}{dE} dE + 2\left(E \frac{dZ}{dE}\right)_{\eta_0}(\eta - \eta_0) + \frac{\pi^2}{3}(kT)^2 \left(\frac{dZ}{dE} + E \frac{d^2Z}{dE^2}\right)_{\eta_0}.$$

Inserting for $\eta - \eta_0$ from (4.52),

$$E(T) = 2\int_{-\infty}^{\eta_0} E\frac{dZ}{dE}\,dE + \frac{\pi^2}{3}(kT)^2\left(\frac{dZ}{dE}\right)_{\eta_0}$$

$$= E(0) + \frac{\pi^2}{3}(kT)^2\left(\frac{dZ}{dE}\right)_{\eta_0}. \tag{4.56}$$

The electronic specific heat is

$$\frac{dE}{dT} = 2\frac{\pi^2}{3}k \cdot kT\left(\frac{dZ}{dE}\right)_{\eta_0}. \tag{4.57}$$

The results (4.56) and (4.57) can easily be interpreted. The constant energy term in (4.56) is just the energy of the lowest state of the system, obtained by filling all states up to the level of the Fermi energy η_0 and leaving the rest vacant. At a finite temperature T the only states whose occupation numbers are different from these at zero temperature are those within an energy region of the order of kT about η_0. The electrons in this region are the only ones which can adjust themselves to the rising temperature, the rest are still 'frozen in'. It is therefore reasonable that the specific heat should be of the order of k for each of those electrons. Their number is the density of states dZ/dE, times the width kT of the energy region.

In a non-metallic crystal η_0 coincides with the upper limit of an energy band, which at this point does not overlap with any other. Hence in that case $(dZ/dE) = 0$, and there is no electronic specific heat, as long as the approximation (4.50) is justified.

In a metal, we may as a crude estimate for orientation expect dZ/dE to be of the order of the number of atoms divided by an energy of the order of the band width. Since this is in general of the order of a few eV, whereas kT at room temperature is of the order of $3 \cdot 10^{-2}$ eV, the electronic specific heat at room temperature is something like 1 per cent. of the Dulong-Petit value.

This result, which was first pointed out by Sommerfeld, removes the discrepancy that existed between the evident fact that metals contained large numbers of apparently free electrons, and the lack of any significant contribution to the specific heat that could be ascribed to them.

However, in an ordinary metal, the electronic specific heat is more easily observable either at rather higher, or at very low temperatures. At higher temperatures, when the vibrational specific heat is approximately constant, a term proportional to the temperature is more easily noticed. However, we have seen in § 2.3 that the crystal lattice itself gives a correction to the specific heat proportional to T, and a precise knowledge of this effect would be needed to isolate the electron contribution.

It is easier to observe the effect at low temperatures when the vibrational specific heat becomes proportional to T^3, so that, in the limit of very low T, the electronic part will be dominant. Experiments in this region therefore give an estimate of the density of electron levels near the Fermi energy.

Our crude estimate of dZ/dE will be unreasonable in two cases: (a) When the Fermi energy η_0 lies in an exceptionally narrow band. This happens in the case of the transition metals like Pd, in which the free atom contains an unsaturated inner shell, so that some conduction electrons are in rather small orbits, for which the overlap integrals of § 4.2 are small. (b) The other exception arises if we are dealing with a band that is very nearly empty, or very nearly filled. In that case the energy is a quadratic function of k near the limiting value, and in that case we have seen that the distribution of states is nearly the same as for free electrons, except for the value of the mass. In that case one easily verifies that

$$\frac{dZ}{dE} = \frac{V}{4\pi^2}\left(\frac{2m^*}{\hbar^2}\right)^{\frac{3}{2}}|E-E_1|^{\frac{1}{2}} = \frac{3m^*}{\hbar^2}\left(\frac{V}{6\pi^2}\right)^{\frac{2}{3}}(Z_1)^{\frac{1}{3}}, \qquad (4.58)$$

where m^* is the effective mass, V the volume, E_1 the energy of the minimum or maximum, and Z_1 the number of electrons in the band or the number of empty places, according to which is small.

In case (a) the density of states is much larger than the crude estimate given previously, and in case (b) it is smaller. In both cases the further derivatives of Z with respect to E will be larger, and the higher terms neglected in (4.50) may no longer be negligible, even at fairly low temperatures. We say in that case that the electron gas is only partly degenerate.

4.7. Surface problems

So far we have taken the crystal as infinitely extended for all practical purposes and used the cyclic boundary condition merely to obtain the correct number of states. In some problems we are, however, concerned with phenomena that specifically involve the surface of the metal, and then a more careful approach is needed. The most important new question is then to place the position of the energy bands absolutely, choosing, for example, the most natural zero of energy, that of an electron at rest outside the metal. At first sight one might think that the potential energy function $V(\mathbf{r})$ could be obtained in principle from a knowledge of the structure of the atomic cores, and that, for example, in the tight binding approximation of § 4.2 a knowledge of the atomic core potential $U(\mathbf{r})$ would be sufficient, so that the determination of the energy bands is then merely a matter of calculation.

The situation is not so simple, however, because, as was pointed out earlier, the atomic cores carry a positive charge and to get realistic results one must allow for the mean negative charge density due to all the conduction electrons. In the inside of the metal this is bound to be distributed uniformly, so that the total charge in a unit cell just cancels that due to the core, and each cell is neutral. But there is no reason to suppose that this exact cancellation is still valid near the surface. To be sure, we expect the crystal as a whole to have no net charge, so that for each of the surface atoms too there will somewhere be a negative charge to compensate the positive charge of the core. We do not know, however, how this will be distributed in space. It may be, for example, that the average charge distribution of the electrons near the surface layer is unsymmetrical and spreads out into space more than it spreads into the inside of the crystal.

We can imagine this state reached by building the crystal first from electrically neutral cells, and then displacing the charge of the surface atoms a little outwards, without moving the cores. The result is an electric double layer, which will lower the potential inside the metal by a constant amount, and will, of course, shift the position of the energy bands by the same amount.

The problem of actually determining this shift is very involved. It could in principle be obtained by the method of self-consistent fields, since the values of the potential near the surface affect the electronic wave functions, and hence the charge distribution of the conduction electrons. The problem is further complicated by the fact that the exact position of the surface atoms is also not known, since the absence of neighbours on one side will disturb the symmetry, so that the ideal lattice points are no longer equilibrium positions. In general, one has to treat the potential shift near the surface as an empirical parameter.

This shift will not only depend on the physical state of the surface, and be rather sensitive to adsorbed layers of other atoms, but it will also be different according to the crystallographic orientation of the surface. For this reason it is not convenient in practice to count the energy from the external zero. It is better to think of the position of the energy bands inside as given in relation to some arbitrary level and then state the value U of the external zero in relation to the same level. Then the position of the energy bands is not sensitive to the state of the surface, and the surface dependence is contained in U.

One possible choice is to choose the bottom of the conduction band as reference level. Then U is the energy needed to move an electron which is at the bottom of the conduction band to a state of rest outside.

At low temperatures there are electrons in all states in the metal up to energy η. The least energy required to remove an electron from the metal at zero temperature is therefore

$$W = U - \eta, \tag{4.59}$$

which is called the 'work function'. The quantities on the right-hand side are dependent on the reference level. W is not, but, like U, it depends on the state of the surface.

Now consider two different metals in contact with each other. The nature of the contact is unimportant as long as electrons are able to move from the one to the other. Then in order to have the whole in statistical equilibrium the electron distribution in both metals must be represented by the Fermi function $f(E)$ with the same value of η. Otherwise electrons are transferred and form a double layer near the contact surface until the condition has been achieved.

Except for the immediate neighbourhood of the surface of contact, each metal still has its normal properties, and conditions at the external surfaces will still be normal. The contact has, however, equalized the Fermi energies η. (4.59) therefore shows that the potentials just outside the two metals differ now by

$$U_1 - U_2 = W_1 - W_2. \tag{4.60}$$

This potential difference is known as contact potential, and according to (4.60) it equals the difference in the work functions. In this sense surfaces of a metal crystal of different crystallographic orientations also have a contact potential relative to each other.

It follows from our discussion of (4.52), which gives the temperature dependence of η, that for two normal metals we would expect the contact potential to be practically independent of temperature.

Next we consider the escape of electrons from the metal at high temperatures. As a preliminary, consider the surface region in equilibrium, neglecting (in this case unrealistically) the effect of the space charge outside the metal which is produced by the electrons. In that case the Fermi distribution must be valid also outside the metal, and, with the same choice of the zero level of energy as previously, the number of electrons in a state of kinetic energy E_{kin} outside the metal is

$$2f(U+E_{\text{kin}}) = \frac{2}{1+e^{(U+E_{\text{kin}}-\eta)/kT}}. \tag{4.61}$$

At all reasonable temperatures the exponent of the exponential is large, and hence the 1 in the denominator may be neglected. Using also the definition (4.59) the expression may be written as

$$2e^{-(W+E_{\text{kin}})/kT}. \tag{4.62}$$

Using the result of statistical mechanics that the number of states of a free particle is
$$(2\pi\hbar)^{-3}\, d\Omega, \tag{4.63}$$

where $d\Omega$ is the volume element in phase space, we can evaluate the density of electrons outside the metal and their velocity distribution (which is, of course, the Maxwell distribution, since the dependence of (4.62) on kinetic energy is just the ordinary Boltzmann function).

These results (apart from the neglect of space charge) apply to thermal equilibrium, or in other words to the condition in which as many electrons are incident on the metal from some external source as are leaving it. It may also be used to give us the rate of emission from the surface if we assume the same distribution for electrons travelling away from the metal surface, but remove the electrons travelling towards the metal. This is justified if the electrons just outside the metal have made their last collision inside. Since the rate of emission is extremely small, if measured in terms of the density of electrons inside the metal and their velocities, the effect of the collisions inside the metal is still the same as in equilibrium. The model depends therefore on the assumption that the electrons are not likely to suffer any disturbance on leaving the metal. Since there is a rise of potential near the surface over a distance comparable to the electron wavelength, there is, in fact, a finite probability of the electron being scattered back at the surface, and this will reduce the emission by a numerical factor which is not very important in practice.

We require, therefore, the flux of electrons from (4.62) and (4.63), counting only electrons whose velocity is directed away from the surface, and an elementary integration gives the themionic emission current per unit area as

$$\frac{em}{2\pi^2\hbar^3}(kT)^2 e^{-W/kT}. \tag{4.64}$$

The exponential factor is well confirmed by experiment, and for clean surfaces the relation of the work function found in this way with the contact potential is in good agreement with observation.

The absolute magnitude of the other factor and its variation as T^2 is much harder to confirm, since it is very sensitive to small errors in the exponent, and to any slight variation, due to secondary causes, of the work function with temperature. In addition, any small inhomogeneity of the surface, due to impurities, geometrical shape, or crystal structure, will complicate the problem. In a microcrystalline surface, in which not all the grains expose crystallographically equivalent faces, (4.64) would be replaced by a combination of such expressions with slightly different work functions.

So far we have neglected space charge. In fact, the escape of a few electrons produces a space charge which raises the potential at some distance from the metal, and thereby prevents the escape of further electrons. For this reason the current from the surface depends on the presence of an external field, and the 'saturation current' (4.64) is obtained only if the field is strong enough to make the space charge effect negligible.

From what was said before about contact potentials it is evident that, in the case of a metal consisting of several layers of different materials, the work function to be used in (4.64) is that of the actual surface material. Hence the practice of covering cathodes with a substance of low work function, which by itself would not have mechanical or electrical properties suitable for use as cathode material.

Even at low temperatures, electron emission is possible if a strong field is applied in a direction to pull electrons away from the surface. In that case, the potential goes down as we go away from the surface, and, at a certain distance x_1 from the surface, the potential energy will have dropped to the value η. This distance is, in fact,

$$x_1 = W/eF \tag{4.65}$$

if F is the field intensity. An electron inside the metal near the edge of the Fermi distribution has therefore enough energy also to move outside the metal at distances greater than x_1 from the surface, but it does not have enough energy to pass over the potential 'barrier' which is formed by the intervening region in which the potential is higher.

Now it is well known that in quantum mechanics a particle has a finite probability of making a transition through such a barrier. This probability contains as the dominant factor an exponential with the exponent

$$\frac{2}{\hbar} \int \{2m(V-E)\}^{\frac{1}{2}} \, dx,$$

where E is the energy of the particle, V the potential energy, and the integral is to extend over the region in which $V > E$. In the present problem $V - E$ is equal to the work function at the surface, and from there decreases linearly to zero at x_1. Hence the exponential factor becomes

$$e^{-\lambda W^{\frac{3}{2}}/F}, \qquad \lambda = \frac{4(2m)^{\frac{1}{2}}}{3e\hbar}. \qquad (4.66)$$

As in the previous case, there are factors depending on the field intensity multiplying the exponential, but again they are hard to isolate in practice because of minor variations of the exponent over the surface and in this case also because any geometrical irregularity of the surface may lead to a local increase of the electric field.

In this discussion of surface effects we have taken into account only the influence of the surface on the position of the main energy bands. In addition, the surface will cause changes in the electron states. One very obvious consequence of the presence of a surface is that, instead of progressive waves, we must introduce standing waves which do not carry any current in a direction normal to the surface. The phase of such a standing wave depends on the behaviour of the potential near the surface and, since this phase influences the electron density near the surface, will itself influence the potential.

A less obvious feature is the fact that in addition to wave functions which change periodically with a displacement by successive multiples of a lattice vector (4.7), we cannot now rule out wave functions which decrease exponentially as we pass from the surface into the metal. In other words, the component of **k** in a direction at right angles to the surface may now be complex. This gives rise to surface waves. Their number is always small compared to that of the periodic solutions, and they are therefore always negligible in statistical arguments concerning the whole metal. For problems in which the surface is of special importance, they may, however, have to be taken into account.

V
COHESIVE FORCES IN METALS

5.1. General discussion

THE energy, both potential and kinetic, of the conduction electrons is comparable with the binding energy of the substance. It must therefore be taken into account in discussing the stability of lattices. We have seen in the last chapter that the energies of the electronic states depend not only on the nature of the attractive forces near each atomic core, but also on the lattice.

The problems to which, in principle, the theory provides the answer are, in this connexion:

(a) The energy of a lattice of given type, as a function of the lattice spacing and other lattice parameters. The energy minimum will then give the binding energy for that particular lattice type; the spacing for which the minimum occurs is the equilibrium spacing, and the curvature of the energy curve determines the compressibility. If one can obtain the same result for a sheared lattice, this gives the other elastic constants.

(b) Comparison of the minimum energy for different lattice types determines the type which has the lowest energy and will therefore occur at low temperatures.

A quantitative attack on these two problems is hopeful only for lattices of a simple type. Certain metals have, as their most stable forms, very complicated and unsymmetric lattices. Here it is in some cases possible to see qualitative reasons for the occurrence and nature of such lattices.

Before going into further detail, it is useful to bear in mind the following simple and qualitative picture:[†]

Consider a crystal built up by bringing isolated atoms together from a large distance. In an isolated atom the energy of each electron is a compromise between the requirements of low potential energy and small kinetic energy. The first favours concentrating the wave function near the potential minimum, but this, by the uncertainty principle, tends to raise the kinetic energy. This compromise determines the size of the region occupied by the electron; it is roughly equal to the region in which

† This picture was, for molecules, first discussed by Hellman (1933).

the potential energy is below the actual energy of the electron, and its linear dimensions are roughly equal also to the wavelength corresponding to the mean kinetic energy.

As the atoms approach, each electron has the opportunity of spreading out into adjacent atoms. This reduces its kinetic energy and allows it to make better use of the attractive potential. This spreading of the wave functions is, however, limited by Pauli's principle. As several electrons are using the same volume, they must be in different states, either as regards spin, or as regards their velocity. If the isolated atoms were already saturated, as in the noble gases like He, then the attractive region is already used by as many electrons of the same energy as the Pauli principle allows. In that case there is no gain from spreading out the wave functions, and only a loss from the fact that the atoms overlap, so that the amount of space available to each electron is now curtailed.

For unsaturated atoms, however, there is a gain, and hence an attraction, which may be regarded as the main cause of binding. If the atoms are pushed too close together, the energy rises again because the volume which is now available to all electrons jointly diminishes further and thus increases their kinetic energy. At the same time the cores begin to overlap and, since they are saturated, increase the repulsion.

In discussing the energy in more detail we shall take, for simplicity, the case of a monovalent metal, with one conduction electron per atom. We must then include the following contributions:

(1) The energy of each electron in the potential field $V(\mathbf{r})$. It is convenient to subdivide this into:

(1 a) the energy of the lowest state in the conduction band, and

(1 b) the spread of energies within the band, up to η.

For this purpose the potential $V(\mathbf{r})$ must be defined precisely. It comprises, in the first place, the electric potential due to the atomic core, i.e. the atomic nucleus and the closed electron shell, excluding the last electron. It is a good approximation to assume the charge distribution in the core to be the same as in a single atom. In addition, we must allow for the mean charge due to all conduction electrons. Strictly speaking, we should exclude here the mean charge of the one electron which we are considering, as in Hartree's treatment of the atom. However, the number of electrons in a metal crystal is so large that the difference of one is negligible.

If we applied this rule consistently, each unit cell would contain zero net charge, and the electron would be taken as moving, not in a field similar to the field acting on the last electron in an atom, but rather like that in a negative ion. This approach over-estimates the probability of finding two conduction electrons in the same cell, which is later corrected by the correlation term. (See below.)

Since, however, the correlation energy cannot be calculated exactly, it is desirable to make this correction small, and for this purpose Wigner proposed using for the potential in each cell the same as that used in the atom, i.e. that due to the core only. Since we assume the potential to be periodic, and repeat it in all cells, this will imply a discontinuity of the electric field on the cell boundary, which implies a surface charge. It is clear that this will nevertheless give reasonable answers in the case of widely separated atoms, where the chance of finding more than one valency electron near any one atom is small.

For a real metal this method might appear somewhat inconsistent, but it is still preferable to the alternative treatment which would leave a large correlation correction to be accounted for.

In addition, we have to consider:

(2) The internal energy and the interaction between the atomic cores.

(3) The correlation between the conduction electrons. Since, up to this point, we have treated the electrons as moving independently, their interaction has been allowed for on an average basis. In fact, the probability of finding two electrons close together is less than on a random distribution. There are two cases.

(3 a) If the spins of two electrons are opposite, the correlation between them is entirely due to the force between them. An exact treatment of this effect would require the solution of a many-body problem. One may estimate it by working it out for free electrons, ignoring the periodic potential for this purpose.

(3 b) If two electrons have parallel spins, the Pauli exclusion principle already produces a correlation between their space coordinates. The antisymmetry of the wave function requires it to vanish when the electrons coincide, and since a very steep rise would lead to a high kinetic energy, it will, in fact, remain small over a finite range of distances, of the order of the wavelength. This effect could be calculated by forming the total wave function as an antisymmetric combination of single-particle wave functions. In practice, however, even the calculation of the mean interaction from this is rather complicated, and a free-electron approximation is used instead. Since in the case of parallel spins the electrons are kept apart by the exclusion principle, their electrostatic interaction is small, and the further correlation caused in this case by the forces is quite a small correction.

From what was said in § 4.7 about double layers at the surface, it might appear that, in defining our problem, we should have to pay attention to conditions at the surface. This is not the case, however, since a change in surface conditions will produce only an electric double layer, which will alter the electric potential in the interior by a constant amount. Since the interior of the metal is electrically neutral, this will not cause any change in the total energy.

5.2. The Wigner–Seitz approximation

For the treatment of (1 a), the bottom of the conduction band, Wigner and Seitz developed a method which is particularly suited to the treatment of wave functions for which $\mathbf{k} = 0$.

Consider two opposite faces of the unit cell of the crystal lattice. The periodicity condition (4.7) means then that both the wave function ψ and its derivative must be equal at corresponding points on these faces. If, in addition, the cell has mirror symmetry about the plane half-way between these faces, it follows also that the derivative normal to the face must be equal and opposite at corresponding points. The two requirements together then make it necessary that the normal derivative of the wave function for $k = 0$ vanishes on these faces, and hence over the whole surface of the unit cell.

Since for the most important types of metallic lattice, namely the cubic body-centred and face-centred and the hexagonal close-packed structure, the boundary of the unit cell is a surface of high symmetry, it is a good approximation to replace it by a sphere of the same volume. We shall call its radius r_0. The distance of the points on the actual boundary differ from r_0 by at most 12 per cent.

If we require, as boundary condition, that the normal derivative of the wave function be zero on the sphere, the problem has spherical symmetry, and the solution will be a wave function which depends only on r, the distance from the centre. Compare this with the wave function of the isolated atom. For this the potential beyond r_0 is small, and the wave function beyond r_0 is of the form

$$\psi_0 = \text{constant} \times \frac{e^{-\lambda r}}{r}; \qquad \lambda = (-2mE_0/\hbar^2)^{\frac{1}{2}}, \tag{5.1}$$

where E_0 is the (negative) electron energy. For the atomic wave function therefore

$$\left(\frac{1}{\psi_0}\frac{\partial \psi_0}{\partial r}\right)_{r_0} = -\left(\lambda + \frac{1}{r_0}\right). \tag{5.2}$$

Now it is well known (and can be confirmed by inspecting the radial

§ 5.2   COHESIVE FORCES IN METALS   105

Schrödinger equation) that by lowering the energy we tend to make the wave function curve away from the r-axis, and it will be seen therefore that a decrease of the energy is required to go from the boundary condition (5.1) to zero slope at r_0. Hence for a reasonably large value of r_0 the energy is decreased. The two functions ψ and ψ_0 are shown qualitatively in Fig. 12. Since the required change in the curvature is spread

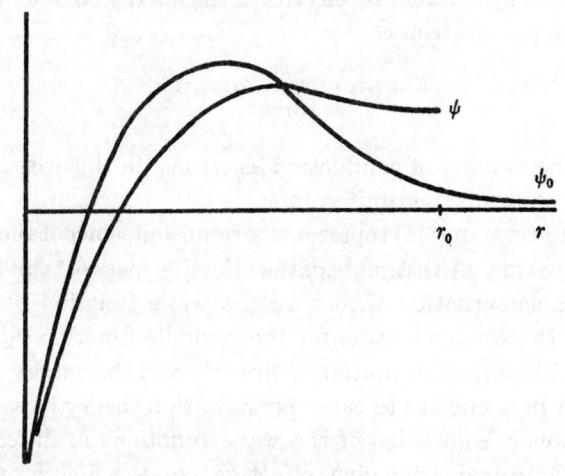

FIG. 12.

out over a large distance, and is small compared to the curvature inside the atomic core, the change in energy is fairly small on the atomic scale. If r_0 is made smaller, the energy changes decrease again and will be zero when r_0 coincides with a maximum of ψ_0.

This energy decrease is evidently connected with the fact that the electron is now capable of spreading out over neighbouring cells, as discussed qualitatively in the preceding section.

The energy of the lowest electron state in the band can thus be obtained easily from the numerical integration of an ordinary differential equation if the potential energy function is known. From the general discussion given above, we expect this energy to decrease as the lattice constant, and with it r_0, is decreased, but to go through a minimum and rise for very small values.

It is very much more difficult to calculate the energies of states for which the wave vector **k** does not vanish, since the problem is no longer isotropic. There exist many calculations which give approximate values for the dependence of the energy on **k**, but they are too complicated to be summarized here.

For small **k**, we have seen in § 4.5 that, for a cubic crystal, the difference $E(\mathbf{k})-E(0)$ is proportional to k^2, so that

$$E(\mathbf{k}) = E(0) + \frac{\hbar^2}{2m^*} k^2, \tag{5.3}$$

where m^* is the effective mass. If this approximation is valid for all states occupied by conduction electrons, the energy spread increases the mean energy per electron to

$$E(0) + \frac{3\hbar^2}{10m^*} (3\pi^2 n)^{\frac{2}{3}}, \tag{5.4}$$

where n is the number of conduction electrons (in this case the same as the number of atoms) per unit volume.

The second term in (5.4) represents a repulsion, since it increases with decreasing spacing. An estimate of the effective mass m^* can be obtained by applying perturbation theory to the wave function of the lowest state, using the wave equation for the periodic function $u_\mathbf{k}$ defined by (4.9). Since this equation contains k linearly and the energy change is of second order in k, one has to carry perturbation theory to second order, which requires a knowledge of the wave functions in different bands. An approximate evaluation gives effective masses which for most monovalent metals are close to the mass of a free electron. This is connected with the fact that the wave function for $\mathbf{k} = 0$ (cf. Fig. 12) is approximately constant over most of the volume of the lattice cell (distances less than $\frac{1}{2}r_0$ contain only one-eighth of the volume of the sphere), and therefore very similar to the wave function of a free electron at rest.

Contribution (2) of the preceding section contains the internal energy of the core which, together with the energy value of the last electron, makes up the energy of an isolated atom. It further contains the electrostatic interaction of different cells which, since the cells are electrically neutral and almost spherical, is negligible. Lastly it contains the repulsion due to the overlap of the core wave functions. This is negligible in the alkalis, which have small cores and large lattice constants, but has a strong effect on the lattice spacing and compressibility of such metals as Cu, Ag, Au.

To discuss item (3), the correlation energy, one must first remember that we have been inconsistent in omitting, within each cell, the potential due to the charge distribution of the conduction electrons within this cell. We must therefore add, in effect, the interaction of such a charge distribution with itself. This is usually done by assuming the charge uniformly spread over the cell. On the other side of the balance we have the reduction of the interaction due to the correlation effect.

If we again treat the electrons as free, the calculation of item (3 b), concerning electrons with parallel spin, is elementary. The antisymmetric wave function for two electrons of wave vectors $\mathbf{k}_1$ and $\mathbf{k}_2$ is

$$\frac{1}{\sqrt{2V}} (e^{i\mathbf{k}_1.\mathbf{r}_1} e^{i\mathbf{k}_2.\mathbf{r}_2} - e^{i\mathbf{k}_2.\mathbf{r}_1} e^{i\mathbf{k}_1.\mathbf{r}_2}), \tag{5.5}$$

where V is the volume. The square of its modulus is the probability of finding the two electrons in the positions $\mathbf{r}_1$, $\mathbf{r}_2$. With the notation

$$\mathbf{r} = \mathbf{r}_1 - \mathbf{r}_2, \tag{5.6}$$

this probability is

$$\frac{1}{V^2}\{1 - R(e^{i\mathbf{k}_1.\mathbf{r}} e^{-i\mathbf{k}_2.\mathbf{r}})\}, \tag{5.7}$$

where the symbol R means 'real part of'. To obtain the total effect, this expression should be averaged over all pairs of electrons with parallel spin, which means averaging the values of $\mathbf{k}_1$ and $\mathbf{k}_2$ over a sphere of radius k_0. An elementary integration gives the result

$$\frac{1-g}{V^2} = \frac{1}{V^2}\left\{1 - \frac{9}{\kappa^6}(\sin\kappa - \kappa\cos\kappa)^2\right\}, \tag{5.8}$$

where

$$\kappa = k_0 r. \tag{5.9}$$

Without the exclusion principle we would only have the first term in (5.8), hence $g(r)$ represents the correlation effect. At small distances $g(r)$ is equal to -1, expressing the fact that electrons of parallel spin are never found at the same point, and g becomes negligible for distances greater than $1/k_0$.

The effect of this correlation on the energy per electron is then (remembering that the number of pairs of electrons with parallel spin is one-quarter of the square of the total number)

$$\tfrac{1}{4} n e^2 \int \frac{g(r)}{r} d^3r = \frac{9\pi n e^2}{k_0^2} \int_0^\infty \frac{d\kappa}{\kappa^5} (\sin\kappa - \kappa\cos\kappa)^2, \tag{5.10}$$

where n is again the electron density.

The integral comes to $\tfrac{1}{4}$, and, inserting for k_0, the reduction of the energy is

$$\tfrac{3}{4} e^2 \left(\frac{3n}{\pi}\right)^{\tfrac{1}{3}}. \tag{5.11}$$

The correlation between electrons of opposite spin is a dynamical problem even for free electrons. It is possible only to obtain a fairly crude estimate which shows this term to be rather less than (5.11), at most about one-quarter of it.

The greatest uncertainty in the method which I have sketched lies in the fact that both the mean density of conduction electrons in a cell and the correlation effects are treated as for free electrons. Since they go in opposite directions, the errors caused by this treatment are likely to compensate to some extent. In fact, the net effect of the interaction within the cell and the correlation as calculated is very small; it must also be small in reality in the limiting case of strong binding, since it then becomes correct to treat each electron as having its own atom and neglect the interaction between each electron and other atoms. It is therefore plausible that also for the case of moderate binding the free-electron treatment does not cause a serious error. If one had followed the apparently more consistent method of taking the electron density into account in defining the cell potential, one would have treated one of the two opposing terms more exactly than the other, with worse results.

It would take us too far to survey the results of such calculations, but they give fair agreement with the observed constants of metals with simple structures.

Similar techniques have been successfully applied to the calculation of the elastic constants, by starting from a slightly sheared lattice. This destroys the approximate spherical symmetry of the unit cell, and both the treatment of the boundary condition for $\mathbf{k} = 0$ and the estimate of the energy spread become more involved.

5.3. Distorted structures. Linear chain

We now turn to the question why the structures of certain metals and alloys are rather more complicated than the simple close-packed or body-centred lattices which, on any picture of direct interatomic forces between identical atoms, would be expected to give the least potential energy.

Some of these structures may, in fact, be visualized by starting from a very simple lattice and introducing a slight displacement of the atoms which decreases the translational or rotational symmetry or both. We shall first discuss the effect of such displacements in one dimension.

Consider the periodic potential of a linear chain of atoms, as in § 4.1. If all the atoms are equidistant, with a distance a, then all multiples of a are lattice vectors, and the basic cell in reciprocal space is, as usual, the interval

$$-\frac{\pi}{a} < k < \frac{\pi}{a}. \tag{5.12}$$

Now suppose we distort the chain by displacing each atom a little, the displacement repeating every rth atom. A simple example would be to move every rth atom by a small amount in the same direction.

This immediately reduces the translational symmetry. We must now

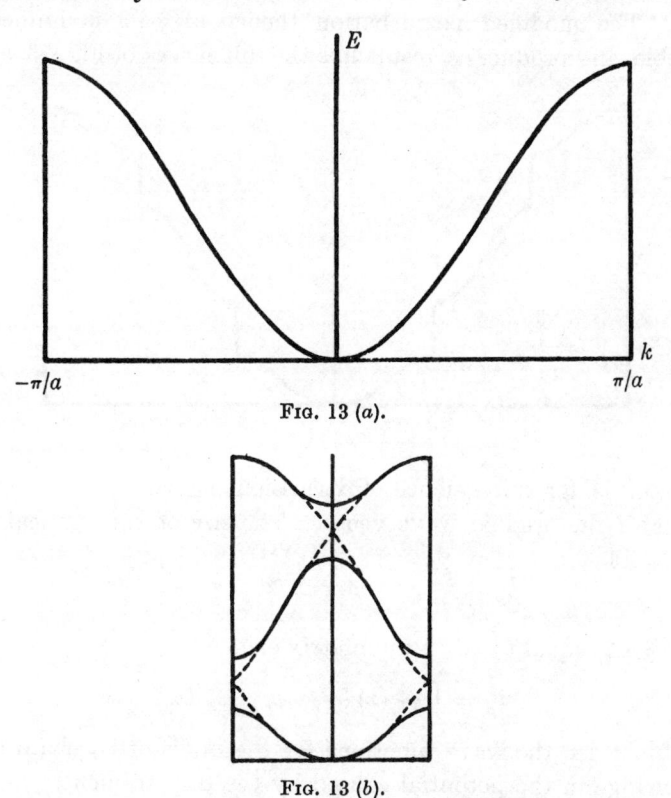

Fig. 13 (a).

Fig. 13 (b).

regard our cell as containing r atoms, and only multiples of ra are lattice vectors. The cell in reciprocal space is now

$$-\frac{\pi}{ra} < k < \frac{\pi}{ra}. \qquad (5.13)$$

The energy curve $E(k)$ for an electron in the potential of the original chain, which might look like Fig. 13 (a), will be modified by the distortion. Because of the reduction in the symmetry, this curve represents now r different bands with the reduced reciprocal cell (5.13). The broken line in Fig. 13 (b) shows this reduction for the case $r = 3$.

The effect of the displacement can now be treated as a perturbation, bearing in mind that the distorted potential has matrix elements connecting states which appear on the same vertical line in Fig. 13 (b). The situation is now precisely that which we have discussed in § 4.3. Again the usual perturbation theory will fail near the ends and the middle of the reduced zone where two of the dotted lines meet, so that there is a non-vanishing matrix element linking two states of nearly the same energy. The modified perturbation theory of § 4.3 is immediately applicable, and produces a result like the full curves of Fig. 13 (b). The

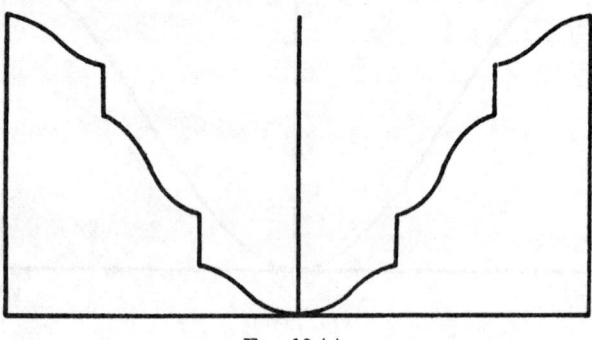

Fig. 13 (c).

same result is, for convenience of discussion, shown again in Fig. 13 (c) in terms of the original wave vector. The size of the vertical breaks occurring at

$$k_\rho = \frac{\rho\pi}{ra}, \qquad \rho = 1, 2, ..., r-1, \tag{5.14}$$

is, as in § 4.3, equal to twice the matrix element

$$V_\rho = \int \psi_{k_\rho}^*(x)\, \delta V(x) \psi_{-k_\rho}(x)\, dx, \tag{5.15}$$

where the ψ are the wave functions for the undistorted chain and δV is the change in the potential caused by the displacement.

The effect of the distortion is to separate any two energy values which in Fig. 13 (b) lie closely above each other. The mean of the two energy values is unchanged, as is evident from (4.38), except for the second-order terms arising from the interaction with more distant states (cf. (4.35)) which are negligible in the present situation.

If, therefore, the states on either side of one of the breaks in Fig. 13 (c) are filled with electrons, the total energy of all electrons will not be affected by the distortion. But if such a break coincides exactly or very nearly with the edge of the Fermi distribution, then the states which are displaced downwards in energy are occupied, and the states which are raised are empty, so that the result is a net reduction of the energy.

It follows that for a one-dimensional metal with a partly filled band the regular chain structure will never be stable, since one can always find a

§ 5.3 COHESIVE FORCES IN METALS 111

distortion with a suitable value of r for which a break will occur at or near the edge of the Fermi distribution.

The gain of energy, and hence the magnitude of the distortion, is greatest when r is a small number, and this means when the number of conduction electrons is in a simple rational relation to the number of atoms. The most favourable case is that in which the Fermi distribution ends at $k = \pi/2a$. Allowing for spin, this means just one electron per atom. In that case $r = 2$, so that every second atom is displaced, or in other words the atoms close up in pairs. This shows a tendency to form a molecular lattice. For any other number of conduction electrons there will be a break at the edge of the Fermi distribution, and thus the band will be split into a number of smaller zones, some of which are filled, and the rest empty.

It is therefore likely that a one-dimensional model could never have metallic properties. However, a complete discussion of this question would have to allow for the fact that, as was stressed in § 1.2, the adiabatic approximation is not valid in the case of a metal, so that the energies calculated with the nuclei at rest cannot be used in physical arguments without great care.

Consider the magnitude of the matrix element (5.15). The displacement u_n of the nth atom must be a periodic function of n with period r, and may therefore be written as

$$u_n = \sum_f A_f e^{ifna}, \qquad (5.16)$$

where the sum extends over values of f which are multiples of $2\pi/ra$, and where $A_{-f} = A_f^*$. Then, if $U(x)$ is again the potential due to a single ion, and if we assume, for simplicity, that, upon displacing the ion, the potential field due to it is displaced without distortion,

$$\delta V(x) \sum_n \{U(x-na-u_n) - U(x-na)\} = -\sum_n u_n \frac{dU(x-na)}{dx} \qquad (5.17)$$

to first order in the displacements. Inserting for u_n from (5.16), we find a sum over the different values of f. One term in this sum gives to the matrix element (5.15) the contribution

$$A_f \sum_n \int dx\, e^{ifna} \frac{dU(x-na)}{dx} \psi_k^*(x) \psi_{-k}(x). \qquad (5.18)$$

Changing the integration variable from x to $x-na$, and remembering the Bloch condition (4.7),

$$A_f \sum_n e^{i(f-2k)na} \int \frac{dU(x)}{dx} \psi_k^*(x) \psi_{-k}(x)\, dx. \qquad (5.19)$$

The integral is now independent of n, and the sum vanishes unless $f-2k$ is a multiple of $2\pi/a$. Since we require the matrix element for the value (5.14) of k, there is always just one value of f for which this is the case.

The coefficient A_f belonging to this is directly related to the X-ray diffraction amplitude for a certain Bragg reflection. According to (3.19), this contains the structure factor

$$\sum_j F_j e^{iq\,d_j}. \tag{5.20}$$

In the case of the distorted lattice the cell contains r identical atoms, and
$$d_j = na + u_j, \qquad j = 1, 2, ..., r.$$

Inserting for u_j from (5.16) and expanding the exponential, the sum (5.20) becomes

$$iF \sum_f A_f \sum_{j=1}^{r} e^{i(q+f)ja}. \tag{5.21}$$

Apart from a constant factor, we may extend the summation over all atoms in the chain, since, for a q value which satisfies the Bragg condition for a cell of length ra, the summand has period ra. For given order of reflection, i.e. given q, only that term contributes for which f differs from $-q$ by a multiple of $2\pi/ra$, so that the sum in (5.19) is proportional to the structure factor (5.20) for reflections in which $q+2k$ is a multiple of $2\pi/ra$. These are, of course, reflections which occur in the distorted chain, but vanish for the undistorted one.

The effectiveness of a certain pattern of displacement can therefore be measured by the extent to which it produces these new X-ray lines.

5.4. Distorted structures. Three dimensions

It is now easy to see that a similar effect can take place in real lattices. It is much less general and much harder to achieve, since a reduction in the translational symmetry will cause energy discontinuities which are planes in **k** space, whereas the boundary of the Fermi distribution in the absence of the distortion is an energy surface, and therefore in general far from plane.

It may happen that an energy surface lies close to such a set of planes. For example, in the simple discussion of § 4.2 of a cubic body-centred lattice with strong binding and only nearest-neighbour interactions, we found that the energy surface which corresponds to half the zone being filled (one electron per atom) was exactly a cube. This cube is the zone boundary for a simple cubic lattice of spacing a. It would therefore be possible to produce energy gaps at this surface, and hence to lower the total energy, by distinguishing the body centres from the cube corners.

This would make the structure one with the translation group of a simple cubic lattice and two atoms per cell. Since, however, the only way to make identical atoms lose their structural equivalence is to displace them, we must necessarily lose the cubic symmetry. We can make all faces of the cube become surfaces of discontinuity by displacing the body centres, relatively to the cube corners, in the direction of a space diagonal. The resulting structure would have orthorhombic symmetry.

This particular structure does not, in fact, occur, the reason being presumably that the assumptions of strong binding and nearest-neighbour interactions are unrealistic.

A very similar distortion does, however, occur in the cases of Bi, Sb, and As, which can be obtained from a simple cubic lattice by displacing every second atom in the direction of a space-diagonal and also altering the angle of the other space-diagonals to the selected one by a few degrees. The position of the Fermi surface is not easily obtained, however, since neither the limit of very strong binding nor that of nearly free electrons is likely to be accurate.

In this case there are five valency electrons per atom (hence ten electrons, or five occupied states, per unit cell) and they must therefore occupy parts of at least six bands. Jones[†] has shown that, in the picture of nearly free electrons, and using momentum space, one can find a region which is bounded by surfaces of discontinuity, and which contains just five states per cell. Part of the planes making up this surface belong to X-ray lines which would be absent in the more symmetric structure, but are prominent for the metals in question. It is reasonable to suppose that the energy surface follows this boundary fairly closely, extending into the exterior near the mid-points of the faces, and leaving the corners vacant.

A similar situation arises in alloys in which certain types of structure occur for specific compositions. The particular concentration ratio for which such a structure is found differs for different pairs of metals. For example, the so-called γ-structure, which is a very complicated cubic structure with fifty-two atoms in the unit cell, occurs near an atomic concentration of 60 per cent. Zn in an Ag—Zn system, near 20 per cent. in Cu—Sn, and near 80 per cent. in Ni—Zn. In all these cases the ways in which the two kinds of atoms are distributed over the sites in the cell are different, but the sites are always very nearly similar. Hume-Rothery pointed out that the composition at which each type of structure occurs corresponds to a particular value of the number of valency

† For further details see Mott and Jones (1936).

electrons per atom. In the case of the γ-structure this ratio is approximately 1·6. This rule can be understood if one assumes, with Jones, that for each type of structure there exists a set of planes of discontinuity which coincide approximately with an energy surface. This structure would then be favoured if the number of electrons is suitable to fill just the region inside that particular surface.

This makes it clear why the nature of the particular atoms and their distribution over the cell does not seem to matter; the main point is that the cell must be of a certain kind of symmetry, and must favour certain X-ray or electron reflections.

These examples give us some insight into the nature of the metallic bond. It is clear that any attempt to represent this in terms of simple forces acting between pairs of atoms would be quite misleading. Apart from the interaction between the ionic cores, which can be so described, the function of the atoms is to provide valency electrons, which then interact collectively with the lattice structure represented by the atoms jointly.

This point of view is now commonly adopted in the discussion of the stability of structures, but it has not yet been applied extensively to dynamical and statistical problems. For example, the statistical mechanics of superlattices in alloys, on which a great deal of mathematical ingenuity has been brought to bear, is usually treated assuming forces only between nearest neighbours, which is hardly realistic. A more accurate treatment would, of course, be very difficult.

Similarly, the problem of the lattice vibration in metals is usually discussed assuming short-range forces between the atoms. In fact, as we saw in § 1.2, it may not be at all possible to define interatomic forces for dynamical purposes, since the effect of the electrons on the motion of the atoms may not be regarded as a static force. This problem is intimately connected with Fröhlich's theory of superconductivity, and we shall return to it in that connexion.

VI
TRANSPORT PHENOMENA

6.1. General considerations. Collision time

ACCORDING to Chapter IV an electron in the field of force of a perfect lattice has stationary states in which the mean velocity and hence the mean transport of charge and of energy do not vanish. This means that we can set up an electric current and an energy flux without an electric field or a temperature gradient to maintain them. In other words, the electric and thermal resistivities are zero for a perfect lattice. The actual resistivities therefore depend on disturbances which, for the equilibrium problems of Chapter IV, are usually negligible. The main sources of such disturbances are (a) lattice vibrations, (b) impurities and lattice imperfections, and (c) the interaction between the electrons. For a good and pure crystal the first of these effects is the most important, and this explains at once the qualitative fact that the resistance of an ideal metal decreases with decreasing temperature, and tends to zero at $T = 0$. This was one of the major difficulties in the classical electron theory of metals.

The problem of treating these disturbances is similar to the calculation of transport phenomena in the kinetic theory of gases, but the nature of the interactions and the importance of quantum effects brings in a number of essentially new points.

The usual approach to the problem is based on calculating the probability, per unit time, of an electron making a transition from one given state to another, and obtaining from this the rate of change with time of the number of electrons in any given state. This rate of change depends on the distribution of the electrons over the various levels, and the stationary distribution is then determined as that for which there is no change with time. In some cases the lattice vibrations will also be affected by the presence of a current. In that case we must also require that the number and distribution of phonons do not change with time, and thus obtain two simultaneous sets of equations which determine both the electron and the phonon distribution.

In using this approach, which is taken over from gas theory, we are making certain assumptions which are sometimes hard to justify. We shall therefore be careful to note these assumptions in our derivation, and we shall later return to the possible refinements that could come from a more rigorous approach.

For preliminary orientation, we shall first obtain the equations which are analogous to the assumption of constant mean free path in gas theory. However, since the electron velocity is not a convenient variable, we shall use the concept of collision time instead of that of mean free path.

The simple picture is then based on the following assumptions:

The electrons suffer only elastic collisions. The probability per unit time of an electron making a collision is $1/\tau$, where τ is called the collision time. This may depend on the energy of the electron, but for given energy is constant (independent of the direction of motion). After the collision the electron is found with equal probability on any part of the energy surface.

Then the average number of electrons making a transition from some state $\mathbf{k}$ to some other state $\mathbf{k}'$ is proportional to the probability of the state $\mathbf{k}$ being occupied, and to the probability of the state $\mathbf{k}'$ being empty (we consider only electrons of a given spin direction), say

$$An(\mathbf{k})\{1-n(\mathbf{k}')\}. \tag{6.1}$$

The probability of the inverse transition from $\mathbf{k}'$ to $\mathbf{k}$ is obtained by interchanging $\mathbf{k}$ and $\mathbf{k}'$, but the coefficient A is the same (law of detailed balancing). The difference between the two rates gives us the net rate of transfer of electrons from $\mathbf{k}$ to $\mathbf{k}'$, which is

$$A[n(\mathbf{k})\{1-n(\mathbf{k}')\}-n(\mathbf{k}')\{1-n(\mathbf{k})\}] = A\{n(\mathbf{k})-n(\mathbf{k}')\}. \tag{6.2}$$

This last result is the same as would have been obtained without the exclusion principle. In the present case we need not consider the exclusion principle in counting collisions.

Consider now all the states with energies between E and $E+dE$. For simplicity of notation we shall assume these all to be part of the same band, and suppress the suffix l, though this is not essential. Their number near an element $d\sigma$ of the energy surface is

$$dz = \left(\frac{L}{2\pi}\right)^3 \frac{d\sigma}{|\text{grad } E|} dE, \tag{6.3}$$

where L^3 is the volume of the crystal, $\text{grad } E$ is the gradient of $E(\mathbf{k})$ in $\mathbf{k}$ space, and $d\sigma$ is the element of area of the energy surface. We write this as

$$dz = \rho \, d\sigma dE, \tag{6.4}$$

where ρ is the surface density of states, which in general varies over the surface, and we call

$$\int \rho \, d\sigma = S(E) = \frac{dZ}{dE}, \tag{6.5}$$

the last expression being the total number of states per unit energy, as in § 4.6.

Then the loss, per unit time, of electrons from the state **k** due to collisions is
$$\frac{1}{\tau} n(\mathbf{k}),$$
whereas the gain from transitions into state **k** is, since these cover all states in the energy shell with equal probability,
$$\frac{\int n(\mathbf{k}')\rho'\, d\sigma'}{\tau S} = \frac{1}{\tau}\bar{n}, \tag{6.6}$$
where $\bar{n}$ is the average number per state on the energy shell. The net rate of change of the number of electrons in state **k** is
$$\frac{dn(\mathbf{k})}{dt} = -\frac{1}{\tau}\{n(\mathbf{k})-\bar{n}\}. \tag{6.7}$$
The solution of this equation in the absence of an external field is
$$n(\mathbf{k},t) = \bar{n}+\{n(\mathbf{k},0)-\bar{n}\}e^{-t/\tau}. \tag{6.8}$$
Any initial deviation from uniformity over the energy shell decreases exponentially with time.

In an electric field in the x-direction the electrons move in k space according to the law (4.42), and hence the rate of change of n due to the field is
$$-\frac{eF}{\hbar}\frac{\partial n(\mathbf{k})}{\partial k_x}. \tag{6.9}$$
Combining the effect of the field with that of the collisions and requiring the electron distribution to be stationary, we obtain the 'Boltzmann equation'
$$\frac{1}{\tau}\{n(\mathbf{k})-\bar{n}\}+\frac{eF}{\hbar}\frac{\partial n(\mathbf{k})}{\partial k_x} = 0. \tag{6.10}$$

The fields that can be maintained in a metal are weak and we may neglect terms proportional to the square of the field intensity. Otherwise we should include the Joule heat, and should not obtain stationary conditions at all unless we also included explicitly the mechanism by which the heat is disposed of. We therefore assume that
$$n(\mathbf{k}) = f(E)+n_1(\mathbf{k}), \tag{6.11}$$
where f is the Fermi distribution and n_1 is small of first order. Then (6.10) becomes, to first order,
$$\frac{1}{\tau}\{n_1(\mathbf{k})-\bar{n}_1\}+\frac{eF}{\hbar}\frac{\partial f}{\partial E}\frac{\partial E}{\partial k_x} = 0, \tag{6.12}$$

or, using the expression (4.41) for the velocity,

$$\frac{1}{\tau}\{n_1(\mathbf{k})-\bar{n}_1\} = -eFv_x\frac{df}{dE}. \tag{6.13}$$

If we average this equation over the energy surface, the left-hand side vanishes identically. The right-hand side also gives zero, since by (4.12) the energy is an even function of $\mathbf{k}$, hence the velocity an odd function, so that its average over the energy surface must vanish.

The constant $\bar{n}_1$ is therefore not determined. This is to be expected since, with only elastic collisions, the total number of electrons of each energy remains unchanged. We may therefore add to n_1 an arbitrary function of the energy. This will not contribute to the current, and is therefore unimportant for the present purpose. One can however show that, if also some inelastic collisions are allowed for, $\bar{n}_1$ has to be zero.

Now the electric current density is, if we choose the volume of the metal to be unity,

$$J_x = 2e\int n_1 v_x \rho\, dE d\sigma = -2e^2F\int \overline{v_x^2}\tau\frac{dZ}{dE}\frac{df}{dE}\,dE. \tag{6.14}$$

The factor 2 allows for the two spin directions, $\overline{v_x^2}$ is the average over the energy surface.

Now df/dE vanishes except in a region of width kT near $E = \eta$, the other factors in the integral vary slowly over that region. We may therefore replace them by their values at η; the integral of df/dE is -1. Then

$$\frac{\partial J_x}{\partial F_x} = 2e^2\left(\tau\overline{v_x^2}\frac{dZ}{dE}\right)_\eta,$$

and, by an obvious generalization to other components, the conductivity tensor $\sigma_{\mu\nu}$ is

$$\sigma_{\mu\nu} = \frac{\partial J_\mu}{\partial F_\nu} = 2e^2\left(\tau\overline{v_\mu v_\nu}\frac{dZ}{dE}\right)_\eta. \tag{6.15}$$

For cubic symmetry this may be replaced by

$$\sigma = \tfrac{2}{3}e^2\left(\tau\overline{v^2}\frac{dZ}{dE}\right)_\eta.$$

In the particular case of free electrons, Z is proportional to $E^{\frac{3}{2}}$; hence $dZ/dE = 3Z/2E$. Using also the fact that $N = 2Z$ is then the total number of electrons per unit volume, and that v^2 is then constant over the energy surface and equal to $2E/m$, we obtain the well-known result

$$\sigma = \frac{e^2 N\tau}{m}. \tag{6.16}$$

Subject to the assumption about the collision time, this result is evidently valid for the electrons near the bottom of a band, if m is replaced by the effective mass, and also for a nearly full band, if N then denotes the number of vacant places.

In the case of spherical symmetry, the work of this section can be extended to cover the case in which the electrons are not uniformly spread over the energy surface after each collision, but in which the probability of a deflexion through a given angle varies with the angle. In that case we may write, instead of (6.7),

$$\frac{dn(\mathbf{k})}{dt} = \int d\Omega' w(\Theta)\{n(\mathbf{k}')-n(\mathbf{k})\}, \qquad (6.17)$$

where the integration is to be taken over a sphere of radius k. $d\Omega$ is the element of solid angle, and Θ is the angle between $\mathbf{k}$ and $\mathbf{k}'$. $w(\Theta)$ is the differential scattering probability.

We expand n and w in spherical harmonics:

$$\left.\begin{aligned} n(\mathbf{k}) &= \sum n_{lm} Y_{lm} \\ w(\Theta) &= \sum \frac{w_l}{2l+1} P_l(\cos\Theta) \end{aligned}\right\}, \qquad (6.18)$$

where the coefficients n_{ml} and w_l may still depend on the magnitude of $\mathbf{k}$. Inserting in (6.17) and using the properties of spherical harmonics,

$$\frac{dn_{lm}}{dt} = 4\pi(w_l-w_0)n_{lm}. \qquad (6.19)$$

A deviation from uniformity in the angular distribution of the electrons still decays exponentially with time if it consists of spherical harmonics of a given order. However, its decay period now depends on the order. We may write

$$\left.\begin{aligned} \frac{dn_{lm}}{dt} &= -\frac{1}{\tau_l} n_{lm} \\ \frac{1}{\tau_l} &= 4\pi(w_0-w_l) = \int \{1-P_l(\cos\Theta)\}w(\Theta)\,d\Omega \end{aligned}\right\}. \qquad (6.20)$$

Now in the isotropic case the right-hand side of (6.13) depends on angle only through v_x, which is proportional to a component of $\mathbf{k}$ and therefore a spherical harmonic of first order, and (6.13) and all subsequent equations are still valid if we replace τ by τ_1, which according to (6.20) gives

$$\frac{1}{\tau_1} = \int (1-\cos\Theta)w(\Theta)\,d\Omega. \qquad (6.21)$$

This shows in particular that the importance of small-angle scattering is proportional to the square of the angle, a result which is valid beyond the isotropic case discussed here.

In the general anisotropic case one may still look for the characteristic deviations which decrease exponentially with time, but in that case the disturbance caused by an external field will be a mixture of different characteristic disturbances, and the discussion becomes more complicated.

The equations of this section may be retained even in this general case, provided we interpret the quantity τ again as a mean value. This will now depend on the variation of the right-hand side of (6.13) over the energy surface, and also on the weight factor used in any averaging process like (6.14) for which the distribution is required.

6.2. Thermal conductivity

If, besides the electric field, there is also a temperature gradient, we obtain in the rate of change of the electron number the further contribution

$$-v_x \frac{\partial n(\mathbf{k})}{\partial x}. \qquad (6.22)$$

The use of distribution functions depending on the coordinates as well as on the wave vector has to be understood in terms of wave packets, in the sense explained in § 2.4. Using again (6.11) and neglecting terms of second order in the temperature gradient, one finds, in place of (6.13),

$$\frac{1}{\tau}\{n_1(\mathbf{k}) - \bar{n}_1\} = -eFv_x \frac{\partial f}{\partial E} - \frac{dT}{dx} v_x \frac{\partial f}{\partial T}. \qquad (6.23)$$

Obtaining again the electric current density as in (6.14) and the energy transport

$$S_x = 2 \int v_x E n_1 \rho\, dE d\sigma, \qquad (6.24)$$

we find the equations

$$\left.\begin{aligned} J_x &= -2 \int \overline{v_x^2}\, \tau\, \frac{dZ}{dE}\left(e^2 F \frac{\partial f}{\partial E} + e\frac{dT}{dx}\frac{\partial f}{\partial T}\right) dE \\ S_x &= -2 \int \overline{v_x^2}\, E\tau\, \frac{dZ}{dE}\left(eF \frac{\partial f}{\partial E} + \frac{dT}{dx}\frac{\partial f}{\partial T}\right) dE \end{aligned}\right\}. \qquad (6.25)$$

Although we have used the collision time concept in deriving these equations, they are, in fact, valid quite generally, as can be seen from the fact that the variation of the two terms on the right-hand side of (6.23) over the energy surface is the same, and that the weight factors used in (6.14) and (6.24) differ only by a function of the energy. Equations (6.25) may therefore be used, with τ having the same meaning in both, as long as we deal only with elastic collisions, so that the electrons of each energy must separately form a stationary distribution.

In evaluating the second terms in (6.25) some care is needed, since the

§ 6.2 TRANSPORT PHENOMENA

integral of $(dZ/dE)(\partial f/\partial T)$ over energy vanishes (the total number of electrons does not change with temperature). The contribution to (6.25) would therefore vanish if the other factors were constant, and it comes entirely from the variation of these factors over the boundary region of f. For these terms the temperature dependent part of (4.50) is therefore essential.

A simple calculation then gives the result

$$\left. \begin{aligned} J_x &= 2e^2 F\left(\tau \overline{v_x^2} \frac{dZ}{dE}\right)_\eta - \frac{2e}{3}\frac{dT}{dx}\pi^2 k^2 T \left(\frac{dZ}{dE}\frac{d}{dE}(\overline{v_x^2}\tau)\right)_\eta \\ S_x &= 2e F\left(\tau E \overline{v_x^2}\frac{dZ}{dE}\right)_\eta - \frac{2}{3}\frac{dT}{dx}\pi^2 k^2 T \left(\frac{dZ}{dE}\frac{d}{dE}(\tau E \overline{v_x^2})\right)_\eta \end{aligned} \right\}. \quad (6.26)$$

For the problem of thermal conductivity we have $J_x = 0$. Eliminating F from this condition, we find

$$S_x = -\frac{2}{3}\frac{dT}{dx}\pi^2 k^2 T \left(\frac{dZ}{dE}\overline{v_x^2}\tau\right)_\eta = -\kappa \frac{dT}{dx}. \quad (6.27)$$

Comparing the thermal conductivity κ with the electric conductivity σ from (6.15)

$$\frac{\kappa}{\sigma} = \frac{\pi^2 k^2 T}{3e^2}. \quad (6.28)$$

This is known as the Wiedemann-Franz law.

This law can also be derived, with a different numerical factor, from a theory using classical, rather than Fermi-Dirac, statistics. It had therefore appeared surprising that the Wiedemann-Franz law was, apart from the factor, well confirmed by experiment, since this looked like a confirmation of the classical statistics, according to which the electron gas obeyed the equipartition law for the specific heat. In fact, from a simple dimensional argument one can see that, for given electron number and collision time, thermal conductivity depends on the product of the mean square velocity and the specific heat per electron. In the case of a degenerate Fermi gas the first is much larger than the classical value $3kT/m$, and the second correspondingly smaller than the classical specific heat, so that the product is about the same.

From the equations (6.25) one can also calculate the thermoelectric coefficients, but one must then include in (6.26) a correction to the first term, since the variation of the integrand over the energy interval near the edge of the Fermi distribution is then of importance.

6.3. Static obstacles. Impurities and imperfections

Turning now to the discussion of the mechanism of the collisions, we discuss first the case in which the cause of the scattering is an irregular static potential. This assumption evidently covers the case of foreign atoms replacing atoms of the lattice or filling interstices, and other faults in the lattice. The effect of the lattice vibrations will require separate consideration.

Let the disturbing potential be $W(\mathbf{r})$ and assume it to be the combined effect of a large number of similar scattering centres, so that

$$W(\mathbf{r}) = \sum w(\mathbf{r}-\mathbf{R}_\nu). \tag{6.29}$$

Assume further that the disturbance caused by each centre is weak enough, and their number small enough, to regard W as a small perturbation.

Then if we write the wave function of an electron at time t as

$$\psi(\mathbf{r},t) = \sum_{\mathbf{k},l} a_{\mathbf{k},l}(t)\psi_{\mathbf{k},l}(\mathbf{r})e^{-iE_l(\mathbf{k})t/\hbar}, \tag{6.30}$$

where the $\psi_{\mathbf{k},l}$ are the solutions for the perfect lattice, in the absence of the disturbance, the wave equation leads to the equation

$$i\hbar\dot{a}_\mathbf{k} = \sum_{\mathbf{k}',l'} (\mathbf{k},l|W|\mathbf{k}',l')e^{i\{E_l(\mathbf{k})-E_{l'}(\mathbf{k}')\}t/\hbar}a_{\mathbf{k}'}, \tag{6.31}$$

where $$(\mathbf{k},l|W|\mathbf{k}',l') = \int \psi^*_{\mathbf{k},l}(\mathbf{r})W(\mathbf{r})\psi_{\mathbf{k}',l'}(\mathbf{r})\,dv \tag{6.32}$$

is the matrix element of the perturbing potential. To first order we may regard the a on the right-hand side of (6.31) as constant; hence

$$a_{\mathbf{k},l}(t) = -\sum_{\mathbf{k}',l'} (\mathbf{k},l|W|\mathbf{k}',l')a_{\mathbf{k}',l'}(0)\frac{e^{i(E-E')t/\hbar}-1}{E-E'}, \tag{6.33}$$

where E and E' will be understood as abbreviations for the energies occurring in the exponent of (6.31).

If, therefore, the electron is known to have been at $t = 0$ in a state $\mathbf{k}, l$, then the probability of finding it in the state $\mathbf{k}', l'$ at time t is

$$|(\mathbf{k}',l'|W|\mathbf{k},l)|^2 2\frac{1-\cos(E'-E)t/\hbar}{(E'-E)^2} = |(\mathbf{k}',l'|W|\mathbf{k},l)|^2 D(t). \tag{6.34}$$

This expression is then, in a sense, the transition probability from state $\mathbf{k}, l$ to $\mathbf{k}', l'$. However, it is strictly valid only if the electron was initially in a single stationary state. Had we taken more than one of the $a(0)$ into account in (6.33), we should have obtained cross products in the square (6.34). Now we shall use such a formula to test the stationary property of the electron distribution, and hence the time $t = 0$ should be an arbitrary instant, after the disturbances have already been acting for some time. It is obvious from (6.33) that, even if at $t = 0$ only one coefficient is non-zero, this will in general no longer be true at time t. This may result in phase relationships between the expansion coefficients

$a_\mathbf{k}$. For example, if the disturbance W consisted of a repulsive potential in some region, the effect would no doubt be to reduce the electron density in that region, and this has to be expressed by means of standing waves, or phase relations between the coefficients of the progressive waves which we have used for the expansion.

If we are to neglect such phase relations, this needs justification. We state therefore

Assumption 1. The nature of the perturbing potential is such that we may obtain the statistical ensemble of electrons by distributing electrons over the stationary states of the perfect lattice with suitable probabilities, but without cross terms.

This assumption is justified if the scattering centres are numerous, and distributed at random, so that in the final probability (6.34) we may average over their positions. Then the square of the diagonal element in (6.34) is, from (6.29),

$$|(\mathbf{k}',l'|W|\mathbf{k},l)|^2 = \sum_{\mu,\nu} |\mathbf{k}',l'|w|\mathbf{k},l)|^2 e^{i(\mathbf{k}-\mathbf{k}')\cdot(\mathbf{R}_\mu-\mathbf{R}_\nu)}. \quad (6.35)$$

If we average over the positions $\mathbf{R}_\nu$ independently, all terms cancel except those with $\mu = \nu$. The effect is therefore proportional to the number of scattering centres, which is reasonable. In addition, if we had taken a cross product instead of (6.34), we would have found the sum

$$e^{i(\mathbf{k}-\mathbf{k}')\cdot\mathbf{R}_\mu - i(\mathbf{k}-\mathbf{k}'')\cdot\mathbf{R}_\nu}, \quad (6.36)$$

which, after averaging over the $\mathbf{R}_\nu$, gives zero, unless $\mathbf{k}' = \mathbf{k}''$. There might still be phase relations between states of the same $\mathbf{k}$ and different l, but, since these will always have different energies, they will not be connected by elastic collisions. We have therefore reduced Assumption 1 to the randomness of the positions of the scattering centres, a requirement closely related to the 'Stosszahl-Ansatz' of the kinetic theory of gases.

Accepting (6.34), the change in the number of electrons in state $\mathbf{k}$, between time 0 and time t, becomes, similarly to (6.2),

$$n(\mathbf{k},l,t)-n(\mathbf{k},l,0) = \sum_{\mathbf{k}',l'} |(\mathbf{k}',l'|W|\mathbf{k},l)|^2 \{n(\mathbf{k}',l')-n(\mathbf{k},l)\}D(t),$$
$$(6.37)$$

where D is the function defined by (6.34).

The function D, considered as a function of E', has a steep maximum at $E' = E$. The width of this maximum is of the order of $\hbar/t$. It is therefore customary to assume that the variation of the other factors in (6.37) with E' is slow, so that we may insert their values for $E' = E$, and carry out the integration on D. Since the integral of D is

$$\int D \, dE = \frac{2\pi t}{\hbar}, \tag{6.38}$$

the number of transitions is then proportional to the time, and we may write (6.37) as

$$\dot{n}(\mathbf{k},l) = \frac{2\pi}{\hbar} \sum_{\mathbf{k}',l'} |(\mathbf{k}',l'|W|\mathbf{k},l)|^2 \{n(\mathbf{k}',l') - n(\mathbf{k},l)\} \delta(E' - E). \tag{6.39}$$

However, in the expression (6.37), the factor $n(\mathbf{k}',l)$ may also be a rather quickly varying function of the energy. This happens when E' lies near η, which is the most important region. In that case, n changes by a large amount over an interval of kT. The approximation which underlies (6.39) is therefore valid only if this is larger than the width of the peak D, i.e. if

$$kT > \frac{\hbar}{t}. \tag{6.40}$$

On the other hand, we cannot make the time interval t too large, since we have used first-order perturbation theory, and this must certainly break down if the time interval is long enough for the electron to have made a second collision. Therefore t must be less than the collision time τ, and therefore we have

Assumption II. $$\frac{\hbar}{\tau} < kT. \tag{6.41}$$

For times shorter than (6.40) we would have to replace the δ function in (6.39) by a wider maximum. Since the δ function expresses the conservation of energy, this looks as if, with small time intervals, the energy might not be conserved. This is not correct. We have to remember that the total energy includes the interaction energy as well as the unperturbed energy $E(\mathbf{k},l)$. It follows from general principles that the total energy is always exactly conserved, but it equals the unperturbed energy only after a long time when the collision is certain to have ceased. In this sense we can interpret $\hbar/kT$ as the duration of a collision, and (6.41) is the condition that the duration of a collision be shorter than the time between successive collisions.

One typical kind of disturbance is the replacement of a few atoms in the lattice by foreign atoms. In that case the use of first-order perturbation theory is not strictly permissible, unless the substituted atoms are very similar to the ones they replace. We should use, instead of (6.39), a more exact expression for the scattering probability of a single centre.

This would not alter the form of (6.39), nor the proportionality of the scattering probability with the number of centres.

The angular distribution of the scattered electrons will be complicated. However, since the size of the scattering centre is of the order of an atomic radius, and the effective electron wavelength $(2\pi/k)$ of the same order, unless we are dealing with a nearly empty or nearly full band, we expect large-angle deflexions to be about as common as small-angle ones, so that it is reasonable to average over the energy surface, and use the collision time concept of § 6.1. A more accurate treatment would involve the solution of an integral equation, but our knowledge of the energy surfaces and of the scattering probability is not usually good enough to warrant this.

The situation is similar when we are dealing with atoms added to the lattice in interstitial positions, or with atoms missing from the lattice. These last two occur always in limited numbers, but the substitution can in suitable cases be carried to a point where an appreciable fraction of the atoms has been changed. If in a crystal of atoms A more and more atoms are replaced by B, we shall eventually approach a pure B lattice (assuming that no change of structure occurs) and then the electrons move again without resistance. This problem was treated by Nordheim (1931), who suggested using as the starting-point a perfect lattice in which the potential was an average of the potentials for either type of atom, weighted with their concentrations. Then it is easy to show that, taking the deviation from this mean as a small perturbation, the scattering probability is proportional to $c(1-c)$ where c is the concentration. This law agrees well with experiment in many cases. Typical exceptions were shown by Mott (1936) to be due to a change in the electron number, and the occupation of the different bands, with concentration.

Other lattice imperfections, such as grain boundaries and dislocations, present greater difficulties for a quantitative treatment.

6.4. Effect of lattice vibrations. General

If the atoms are displaced from their ideal lattice positions, the potential $V(\mathbf{r})$ will be changed by an amount

$$W(\mathbf{r}) = \sum_{\mathbf{n},j} \mathbf{W}_j(\mathbf{r}-\mathbf{a_n}) \cdot \mathbf{u}_{\mathbf{n},j}. \tag{6.42}$$

Here $\mathbf{u}_{\mathbf{n},j}$ is the displacement of the atom $\mathbf{n}$, j, as in Chapter I, $\mathbf{W}_j$ measures the effect on the potential of an infinitesimal displacement of the atom, and depends, as shown, only on the position of the point $\mathbf{r}$

relative to the cell **n**. Terms of second and higher order in the displacements have been neglected.

It is convenient to express the displacements in terms of normal coordinates. Using (1.30),

$$W(\mathbf{r}) = \sum_{\mathbf{f},s} \sum_{\mathbf{n},j} \dot{W}_j(\mathbf{r}-\mathbf{a_n}) \cdot \mathbf{v}_j(\mathbf{f},s) e^{i\mathbf{f}\cdot\mathbf{a_n}} q(\mathbf{f},s). \qquad (6.43)$$

The normal coordinate $q(\mathbf{f},s)$ is known to have matrix elements only between states differing by one unit in the quantum number $N(\mathbf{f},s)$. Hence (6.43) can only cause transitions in which the electron is scattered and at the same time a phonon absorbed or emitted.

We consider, for definiteness, the case of absorption. Using the expression (1.63) for the matrix element of the normal coordinate, the matrix element of W for a transition from **k** to **k**′, and absorption of a phonon in the mode **f**, s is

$$(\mathbf{k}', l' | W | \mathbf{k}, l; \mathbf{f}, s)$$

$$= \sum_{\mathbf{n},j} \int d^3\mathbf{r}\ \psi^*_{\mathbf{k}'l'}(\mathbf{r}) W_j(\mathbf{r}-\mathbf{a_n}) \psi_{\mathbf{k}l}(\mathbf{r}) \cdot \mathbf{v}_j(\mathbf{f},s) e^{i\mathbf{f}\cdot\mathbf{a_n}} \left(\frac{\hbar N(\mathbf{f},s)}{2M^{(N)}\omega(\mathbf{f},s)}\right)^{\frac{1}{2}}. \quad (6.44)$$

Changing the origin in the nth term to $\mathbf{a_n}$, and remembering the Bloch theorem (4.7), this becomes

$$\sum_j \int d^3\mathbf{r}\ \psi^*_{\mathbf{k}'l'}(\mathbf{r}) W_j(\mathbf{r}) \psi_{\mathbf{k}l}(\mathbf{r}) \cdot \mathbf{v}_j(\mathbf{f},s) \left(\frac{\hbar N(\mathbf{f},s)}{2M^{(N)}\omega(\mathbf{f},s)}\right)^{\frac{1}{2}} \sum_n e^{i(\mathbf{f}+\mathbf{k}-\mathbf{k}')\cdot\mathbf{a_n}}.$$

$$(6.45)$$

The last sum again vanishes, unless

$$\mathbf{f}+\mathbf{k}-\mathbf{k}' = \mathbf{K} \qquad (6.46)$$

is a vector of the reciprocal lattice. In collisions between electrons and phonons, the wave vector is conserved, apart from vectors in the reciprocal lattice. Because of our restrictions of **f** and **k** to the basic cell, there is only one possible **f** for given **k**, **k**′, and only one possible **k**′ for given **k** and **f**.

We now proceed, as in the derivation of (6.39), to find the transition probability per unit time:

$$|(\mathbf{k}', l' | A | \mathbf{k}, l; s)|^2 \delta\{E_{l'}(\mathbf{k}') - E_l(\mathbf{k}) - \hbar\omega(\mathbf{f},s)\} |N(\mathbf{f},s)|, \qquad (6.47)$$

where

$$(\mathbf{k}', l' | A | \mathbf{k}, l; s)$$

$$= \sqrt{\left(\frac{2\pi}{\hbar}\right)} \sum_j N \int \psi^*_{\mathbf{k}'l'}(\mathbf{r}) W_j(\mathbf{r}) \psi_{\mathbf{k}l}(\mathbf{r})\ d^3\mathbf{r} \cdot \mathbf{v}_j(\mathbf{f},s) \left(\frac{\hbar}{2M^{(N)}\omega(\mathbf{f},s)}\right)^{\frac{1}{2}}, \quad (6.48)$$

and where **f** is defined by (6.46). The same expression gives the probability of emission of a phonon with wave vector $-\mathbf{f}$, provided we change the sign of ω in the argument of the δ-function in (6.47) and replace $N(\mathbf{f}, s)$ by $N(-\mathbf{f}, s)+1$.

We can now use these results in a Boltzmann equation for the electron distribution, provided we are again prepared to accept Assumptions I and II of the preceding section. Then

$$\dot{n}(\mathbf{k}, l) = \sum_{\mathbf{k}', l', \sigma} |(\mathbf{k}', l'|A|\mathbf{k}, l, \sigma)|^2 \delta\{E_{l'}(\mathbf{k}') - E_l(\mathbf{k}) - \hbar\omega(\mathbf{f}, \sigma)\} \times$$
$$\times [|N(\mathbf{f}, \sigma) + 1|n(\mathbf{k}', l')\{1 - n(\mathbf{k}, l)\} - N(\mathbf{f}, \sigma)n(\mathbf{k}, l)\{1 - n(\mathbf{k}', l')\}]. \quad (6.49)$$

Here we have again used the device introduced in (1.33) and (1.64) of separating positive and negative frequencies. The terms shown explicitly in (6.49) represent, for positive σ, the absorption of a phonon $\mathbf{f}$, s and the transition of the electron from $\mathbf{k}$ to $\mathbf{k}'$ and the inverse process. Negative σ means the emission of a phonon of wave number $-\mathbf{f}$ substituted for the absorption of $\mathbf{f}$, or vice versa. For given $\mathbf{k}$, $\mathbf{k}'$ the δ-function will at most allow one or the other of these cases to be realized.

The change in the distribution of phonons is similarly given by

$$\dot{N}(\mathbf{f}, s) = 2 \sum_{\mathbf{k}, l, l'} |(\mathbf{k}', l'|A|\mathbf{k}, l, s)|^2 \delta\{E_l(\mathbf{k}') - E_l(\mathbf{k}) - \hbar\omega(\mathbf{f}', s)\} \times$$
$$\times [\{N(\mathbf{f}, s) + 1\}n(\mathbf{k}', l')\{1 - n(\mathbf{k}, l)\} - N(\mathbf{f}, s)\{1 - n(\mathbf{k}', l')\}n(\mathbf{k}, l)], \quad (6.50)$$

where, of course, $\mathbf{k}$, $\mathbf{k}'$, and $\mathbf{f}$ are again assumed related by (6.46).

The expressions (6.49) and (6.50) vanish if we insert for the electron number the Fermi function $f(E)$, and for the phonon number the thermal mean value $N^0(\mathbf{f}, s)$ from (2.4). This follows because

$$\frac{N^0(\mathbf{f}, s)}{N^0(\mathbf{f}, s) + 1} = e^{-\hbar\omega(\mathbf{f}, s)/kT}; \qquad \frac{f(E)}{1 - f(E)} = e^{-(E-\eta)/kT}. \quad (6.51)$$

One sees easily that the last bracket in (6.49) and in (6.50) is a difference between two terms whose ratio is unity by (6.51) whenever the energy condition is satisfied so that the δ-function does not vanish.

Under the influence of the collision the Fermi distribution for the electrons and the Planck distribution for the phonons therefore are seen to be stationary. This is, of course, ensured generally by the principles of statistical mechanics.

Consider now a small deviation from statistical equilibrium, which we shall take in the form

$$\left. \begin{aligned} n(\mathbf{k}, l) &= f(E) - g(\mathbf{k}, l)\frac{df(E)}{dE} = f(E) + \frac{1}{kT}\frac{g}{(1 + e^\epsilon)(1 + e^{-\epsilon})} \\ N(\mathbf{f}, s) &= N^0(\hbar\omega) - G(\mathbf{f}, s)\frac{dN^0}{d(\hbar\omega)} = N^0 + \frac{1}{kT}\frac{G}{(e^\gamma - 1)(1 - e^{-\gamma})} \end{aligned} \right\} \quad (6.52)$$

Here we have used the abbreviations

$$\epsilon = \frac{E-\eta}{kT}, \qquad \gamma = \frac{\hbar\omega}{kT}. \tag{6.53}$$

Inserting in (6.49) and (6.50) we then obtain, to first order in g and G, $\dot{n}(\mathbf{k}, l)$

$$= \frac{1}{kT} \sum_{\mathbf{k}', l', \sigma} |(\mathbf{k}', l'|A|\mathbf{k}, l, \sigma)|^2 \, \delta(E' - E - \hbar\omega) \frac{g' - g - G}{(e^{\epsilon}+1)(e^{-\epsilon'}+1)|e^{\gamma}-1|};$$
$$\tag{6.54}$$

$$\dot{N}(\mathbf{f}, s) = \frac{2}{kT} \sum_{\mathbf{k}, l, l'} |(\mathbf{k}', l'|A|\mathbf{k}, l, s)|^2 \, \delta(E' - E - \hbar\omega) \frac{g' - g - G}{(e^{\epsilon}+1)(e^{-\epsilon'}+1)|e^{\gamma}-1|}.$$
$$\tag{6.55}$$

Here the arguments of E, g, etc., are not shown explicitly, but it will be obvious from the context that, for example, g means $g(\mathbf{k}, l)$ and g' means $g(\mathbf{k}', l')$.

We see now that the distributions (6.52) are still stationary if we choose

$$g = \text{constant}, \qquad G = 0 \tag{6.56}$$

or
$$g = \lambda E, \qquad G = \lambda \hbar\omega, \tag{6.57}$$

where λ is an infinitesimal constant. It is indeed evident from (6.52) that the first of these solutions amounts to an infinitesimal change in η, and hence to a change in the total number of electrons, and the second is an infinitesimal change in the temperature. These two parameters are, of course, connected with the conservation of the number of electrons and the conservation of energy. If the collisions were limited in such a way that the vector $\mathbf{K}$ in (6.46) were always zero, as would be the case for free electrons interacting with an elastic continuum, then we should have a further solution in which

$$g = \lambda k_x, \qquad G = \lambda f_x, \tag{6.58}$$

which then arises from the conservation of the total wave vector. It is obvious that the distribution modified by (6.58) carries a current, so that in those circumstances a current could persist in the absence of an electric field; we should have infinite conductivity. This shows that the presence of processes with $\mathbf{K} \neq 0$, or else of disturbances other than the electron-phonon collisions, is of vital importance for the electric resistance.

The situation is very similar to that in the thermal conductivity of non-metallic crystals, discussed in § 2.4.

Fig. 14 illustrates the possible collisions in a one-dimensional case. The full line indicates the $E(k)$ curve for one band, and the broken lines

§ 6.4   TRANSPORT PHENOMENA   129

show the frequency spectrum $\hbar\omega(f)$ for given polarization, plotted with some point on the electron energy curve as origin. Evidently an intersection of the two curves represents a solution of (6.46) and of the energy equation. The diagram is not to scale since, in most metals, the phonon

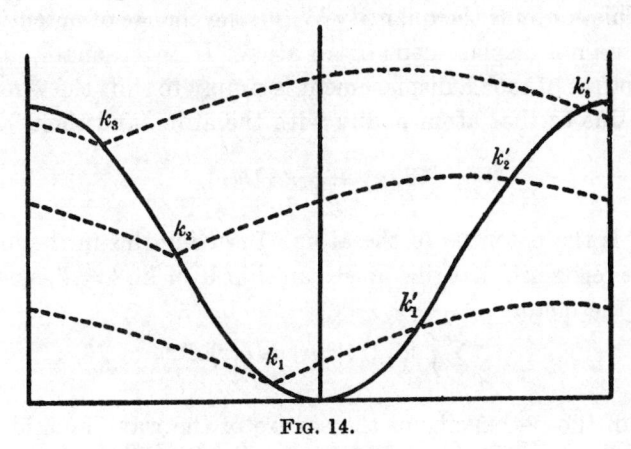

FIG. 14.

energies are much smaller in relation to electron energies than is convenient for a diagram.

Three cases are shown. If the electron is initially at k_1, it can make a transition to k_1', and in that case the wave vector condition (6.46) is satisfied with $\mathbf{K} = 0$. In the cases labelled k_2 and k_3, the difference $k' - k$ exceeds the limits of the basic cell, and in both cases

$$f = k' - k - \frac{2\pi}{a}.$$

In the third case, in particular, f is small. Remembering that the interval for k was chosen purely conventionally, it is more convenient to represent this last case in a diagram like Fig. 14 (a), in which the upper end of the previous figure has been re-plotted in terms of wave vectors lying between 0 and $2\pi/a$. This representation, which is particularly appropriate when we are dealing with a nearly full band, is quite equivalent to the previous one, but the reciprocal lattice vector $\mathbf{K}$ in (6.46) takes

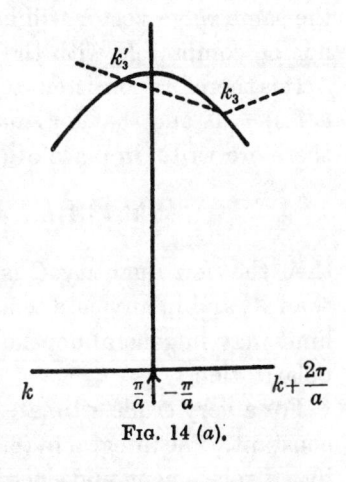

FIG. 14 (a).

in each case a different value. If it should happen that there are no collisions which require a non-zero $\mathbf{K}$ in this representation (or in any other definition of the k) we should still obtain an infinite conductivity. The stationary non-equilibrium distribution would then, of course, look like (6.58) in the new representation.

To conclude this section we require an estimate of the matrix element (6.48). This contains the quantity $\mathbf{W}_j(\mathbf{r})$, the change of potential due to an infinitesimal displacement of an atom. It is reasonable to assume that the effect of such a displacement is simply to shift the whole potential field due to that atom bodily with the atom, and then

$$\mathbf{W}_j(\mathbf{r}) = \operatorname{grad} U_j(\mathbf{r}), \qquad (6.59)$$

where U_j is the potential of the atom. Inserting this in the integral in (6.48), we recognize that this must vanish if $\mathbf{k} = \mathbf{k}'$, $l = l'$, since it then contains the factor

$$\sum_j \int |\psi_{\mathbf{k}l}(\mathbf{r})|^2 \operatorname{grad} U_j(\mathbf{r}) \, d^3\mathbf{r}. \qquad (6.60)$$

Because of the periodicity of the square of the wave functions (Bloch theorem), we may replace here the potential of one cell by the average over all cells, and (6.60) represents the change of energy of the electron upon an infinitesimal displacement of the whole lattice, which, of course, must be zero. Hence, for $l = l'$, the integral in (6.48) vanishes in the limit of long acoustic waves, for which with $\mathbf{f} = 0$, also $\omega(\mathbf{f}, s) = 0$. This will be of importance later. For $l \neq l'$, i.e. transitions from one band to another, the case $\omega = 0$ does not arise, since two states in different bands with the same wave vector will have different energies so that $\omega = 0$ would not be compatible with the energy equation.

It is therefore consistent to assume that the integral vanishes whenever $\omega(\mathbf{f}, s) = 0$, and that for small frequency it is proportional to ω. If we therefore write, in place of (6.48),

$$(\mathbf{k}', l' | A | \mathbf{k}, l, s) = (\mathbf{k}', l' | C | \mathbf{k}, l, s) \sqrt{\{\omega(\mathbf{f}, s)\}}, \qquad (6.61)$$

then the new quantity C is likely to vary less in order of magnitude than A, and in any case tends to a constant limit as $\omega \to 0$, though this limit may in general depend on the direction of $\mathbf{f}$, the state $\mathbf{k}$, and the polarization s.

For a very crude estimate of order of magnitude, we may regard C as constant. The integral in (6.48) is of the dimension of an energy divided by a displacement, and since it represents only the effect of displacing the atoms in one cell out of N, we may expect it to be of order $D\omega/Na\omega_0$, where a is an atomic distance, and D an electronic energy of the order of the band width. ω_0 is the maximum vibration frequency, as in § 2.1. This makes C have the order of magnitude

$$C \sim \frac{D}{a\omega_0\sqrt{M^{(N)}}}.\tag{6.62}$$

6.5. Collisions between electrons

In the classical electron theory of metals the effect of collisions between the electrons was completely ignored on the grounds that in a collision between two free electrons the momentum is conserved, and, since the electric current is proportional to the resultant electron momentum, such collisions also do not alter the current. However, this argument is valid only for free electrons, and for those we have seen in the last section that we could use almost the same reasoning for the collisions with lattice waves. Leaving the lattice structure out of account, collisions with electrons or with elastic waves are insufficient to produce a resistance; allowing for the lattice structure, both will contribute.

However, as we shall see, the electron collisions are in most cases quantitatively less important than the electron-phonon collisions, except perhaps at very low temperatures.

The matrix element of the electron interaction for a transition in which one electron changes from a state $\mathbf{k}_1$ to $\mathbf{k}_1'$ and another from $\mathbf{k}_2$ to $\mathbf{k}_2'$ (we omit the suffix l here for simplicity) is

$$\int\int \psi^*_{\mathbf{k}_1'}(\mathbf{r}_1)\psi_{\mathbf{k}_1}(\mathbf{r}_1)\frac{e^2}{|\mathbf{r}_1-\mathbf{r}_2|}\psi^*_{\mathbf{k}_2'}(\mathbf{r}_2)\psi_{\mathbf{k}_2'}(\mathbf{r}_2)\,d^3\mathbf{r}_1\,d^3\mathbf{r}_2.\tag{6.63}$$

It follows again from the Bloch theorem that this matrix element vanishes unless

$$\mathbf{k}_1'+\mathbf{k}_2'-\mathbf{k}_1-\mathbf{k}_2 = \mathbf{K}.\tag{6.64}$$

For free electrons, when the wave functions are plane waves, $\mathbf{K}$ must vanish, and the matrix element then becomes

$$\frac{4\pi e^2}{L^3(\mathbf{k}_1'-\mathbf{k}_1)^2},\tag{6.65}$$

where L^3 is the volume.

Strictly speaking, (6.65) is too large for small momentum difference, since it is a bad approximation to assume Coulomb's law to hold at very large distances. In a pure two-body problem this would be correct, but, if we remember that there are other electrons present besides the ones under consideration, and that they will neutralize the field of each electron at large distances, we see that it would be more realistic to take a screened Coulomb field in (6.63) and thus obtain a matrix element less strongly divergent than (6.65).

If the electrons are not free, a similar correction has to be applied, and we should also allow for the fact that the mean interaction between the electrons has already been allowed for in the definition of the wave functions. A precise definition of the matrix element is therefore not simple. However, it is reasonable to take its order of magnitude to be roughly equal to (6.65).

Denoting $\sqrt{(2\pi/\hbar)}$ times the matrix element (6.63) by

$$(\mathbf{k}_1', \mathbf{k}_2'|I|\mathbf{k}_1, \mathbf{k}_2),$$

the electron collisions contribute to the rate of change of the electron number a term

$$\dot{n}(\mathbf{k}) = 2 \sum_{\mathbf{k}',\mathbf{k}_2} |(\mathbf{k}', \mathbf{k}_2'|I|\mathbf{k}, \mathbf{k}_2)|^2 \delta(E+E_2-E'-E_2') \times$$
$$\times \{n'n_2'(1-n)(1-n_2) - nn_2(1-n')(1-n_2')\}, \quad (6.66)$$

where the arguments of n and E are indicated by the accent or suffix, and where $\mathbf{k}_2'$ is defined by (6.64). It is again obvious that (6.66) vanishes if $n(\mathbf{k})$ is the Fermi distribution. For a small deviation of the form (6.52) we find

$$\dot{n}(\mathbf{k}) = \frac{2}{kT} \sum_{\mathbf{k}',\mathbf{k}_2} |(\mathbf{k}', \mathbf{k}_2'|I|\mathbf{k}, \mathbf{k}_2)|^2 \delta(E+E_2-E'-E_2') \times$$
$$\times \frac{g'+g_2'-g-g_2}{(e^\epsilon+1)(e^{\epsilon_2}+1)(e^{-\epsilon'}+1)(e^{-\epsilon_2'}+1)}. \quad (6.67)$$

It is of interest to estimate the frequency of collisions for each electron. We choose some state $\mathbf{k}$ in the border region of the Fermi distribution, i.e. in the region in which $E-\eta$ is at most of the order of kT.

The number of collisions suffered by this electron is then given by the contribution to (6.66) of the second term in the bracket, putting $n = 1$. Now n_2 vanishes if E_2 lies above the border region, and $(1-n')$ and $(1-n_2')$ vanish if E' or E_2' lie below the border region. Hence all four energies must, in fact, lie within the border region.

Select, therefore, a value of $\mathbf{k}_2$ in the border region. This can be done in

$$kT\left(\frac{dZ}{dE}\right)_\eta \quad (6.68)$$

ways. The number of ways of choosing $\mathbf{k}'$ is now less than (6.68), because it is further restricted by the energy equation. If we let $\mathbf{k}'$ vary over all the possible (6.68) values, the energy difference between initial and final states would range over some energy interval D, comparable with the width of an energy band. The δ-function will pick out from this region only one particular value for the energy change (namely, zero) and this brings in a factor of the order of D^{-1}. Hence the number of terms that contribute is roughly

$$\frac{1}{D}(kT)^2\left(\frac{dZ}{dE}\right)^2_\eta. \tag{6.69}$$

Using the estimate (6.65) for the matrix element, choosing unit volume, and putting for the difference of wave vectors in the denominator a mean value of the order π/a, we find for the number of collisions per unit time, apart from numerical factors:

$$\frac{e^4 a^4}{\hbar D \pi^4}(kT)^2\left(\frac{dZ}{dE}\right)^2_\eta. \tag{6.70}$$

dZ/dE may be estimated as the electron density, which for a monovalent metal is of the order of a^{-3}, divided by an electronic energy of the order of the band width D. Hence the collision frequency is

$$\frac{1}{\pi^4}\left(\frac{e^2}{a}\right)^2 \frac{(kT)^2}{\hbar D^3}. \tag{6.71}$$

With a of the order of 2.10^{-8} cm., and $D = 2$ eV, this comes to 5.10^{11} sec.$^{-1}$ at room temperature. We shall see later that this is rather less than the effective frequency of collisions due to other sources.

6.6. Collisions at high temperatures

If the temperature is well above the Debye temperature Θ, so that kT is large compared to $\hbar\omega$ for all elastic waves, the problem simplifies considerably. Firstly, it is then in general true that the collisions between phonons are frequent enough to keep the phonon distribution in equilibrium, as we shall verify later. We may therefore put $G = 0$ in (6.54) and ignore equation (6.55). Moreover, the phonon energy is always small compared to the electron energies, and we may regard the collisions as elastic. We may also treat γ as a small quantity in (6.54) so that we may replace ϵ' by ϵ and $e^\gamma - 1$ by γ. If we then allow for the effect of the external electric field (cf. (5.9)), and insert for the matrix element from (6.61), we find

$$\frac{2}{\hbar}\sum_{\mathbf{k}',l',s}|C|^2\,\delta(E'-E)(g'-g)kT = eFv_x. \tag{6.72}$$

The factor 2 arises because we have combined the absorption and emission, so that the polarization index s now takes only positive values.

(6.72) is an integral equation over the energy surface of the kind discussed in § 6.1. We saw there, in particular, that the Wiedemann–Franz law and other consequences of collision-time theory were valid provided all collisions could be regarded as elastic.

We see, moreover, that the temperature enters only through the factor kT on the left-hand side of equation (6.72) and that therefore the solution $g(\mathbf{k})$ of this integral equation is, for given field, inversely proportional to the temperature. This is a well-known empirical law for pure metals at high temperatures.

If impurities and imperfections are not negligible, we must allow for collisions with them on the left-hand side of the equation. The resistances due to either type of collision will be additive if the function g which solves the equation for either type of collision is the same. This is exactly true if the collision time concept is applicable to both types of collision, or in the case of isotropy. In general, deviations from this additivity are not likely to be large.

To estimate the order of magnitude of the conductivity, we take C to be a constant in (6.72). The integral equation is then solved by

$$g = eFv_x\tau,$$

which then makes the term in g' vanish by symmetry. The collision time τ is given by

$$\frac{1}{\tau} = \frac{2}{\hbar}kT\frac{dZ}{dE}|C|^2. \tag{6.73}$$

With the estimate (6.62) for C this becomes

$$\frac{2D^2kT}{\hbar a^2\omega_0^2\rho}\frac{dZ}{dE},$$

where, for unit volume, we have replaced the mass $M^{(N)}$ of the metal by the density ρ. In order of magnitude, dZ/dE is Z/D, and ρ/Z is the atomic mass M. Since $Ma^2\omega_0^2$ is the potential energy for displacing an atom by a distance a, this is again of the order of an electronic energy, i.e. of D. This gives finally, and crudely,

$$\frac{1}{\tau} \sim \frac{kT}{\hbar}. \tag{6.74}$$

This gives the order of magnitude of the conductivity of a normal metal correctly. It also shows that the electron collisions are unimportant if we accept the estimate (6.71).

It remains to justify the assumption that the lattice vibrations are kept in equilibrium by the phonon interaction. This means that a phonon collides more frequently with another phonon than with electrons.

We may estimate the mean collision time for a phonon from (6.50) in the way in which we derived (6.73) and the result is

$$\frac{1}{\tau_{pe}} \sim 2|C|^2 \omega k T \left(\frac{dZ}{dE}\right)_\eta \frac{1}{D}, \qquad (6.75)$$

and hence, from (6.73), $\qquad \dfrac{\tau_{pe}}{\tau_e} \sim \dfrac{D}{\hbar\omega}. \qquad (6.76)$

With $\tau_e \sim 10^{-13}$, we therefore find for τ_{pe} a mean value of the order 10^{-11}. This should be compared with the collision time for phonons colliding with each other, which can be estimated from the thermal conductivity of non-metallic crystals, and, at room temperature, is also about 10^{-13} sec. It is therefore reasonable to suppose the phonons to be in equilibrium.

6.7. Low temperatures

The problem becomes much more involved when the temperature is comparable to the Debye temperature Θ, and we shall discuss only the limiting case when $T \ll \theta$. We may then no longer neglect the energy transfer in collisions, since, for the important collisions, this will turn out to be of the order of kT. The only contributions to (6.54) and (6.55) will now come from small values of **f** for which $\hbar\omega$ is at most of the order kT. If ω is larger, then the absorption of such phonons does not occur because their number is negligible, and the emission does not occur because no electron can lose so much energy. This explains, as was first shown by Bloch, that the resistance of an ideal metal tends to zero at $T = 0$, in spite of the zero-point motion of the lattice, which is connected with the possibility of emitting phonons. In this respect the position differs radically from that in the scattering of X-rays (Chapter III), where we found the zero-point motion to give a finite random scattering even at $T = 0$. This difference arises because in the conduction problem the electrons themselves are nearly in equilibrium at a low temperature, so that processes with large energy transfer to the lattice are ruled out.

We can now estimate the collision time as follows:

For a given electron state **k**, the energy equation may be written, using also the wave-vector condition (6.46) with $\mathbf{K} = 0$,

$$E(\mathbf{k}+\mathbf{f}) - E(\mathbf{k}) - \hbar\omega(\mathbf{f}, s) = 0,$$

or, to first order, $\qquad \hbar\mathbf{f}\cdot\mathbf{v}(\mathbf{k}) - \hbar\omega(\mathbf{f}, s) = 0. \qquad (6.77)$

If we disregard the energy condition, the number of values of **f** with magnitude between f and $f+df$ is

$$\left(\frac{L}{2\pi}\right)^3 4\pi f^2\, df.$$

These, however, will in general violate the energy condition (6.77) by an amount of the order $\hbar f v$, where v is the electron velocity, and therefore the effect of the δ function in (6.49) is to reduce the contribution to

$$\left(\frac{L}{2\pi}\right)^3 \frac{4\pi}{v} f\, df.$$

Since by (6.61) the square of the matrix element contains a factor ω which, for small f, is also proportional to f, we see then that the contribution to (6.49) from the part in which phonons of wave vectors up to f are included, is proportional to f^3. Now collisions in which $\hbar\omega(\mathbf{f}, s)$ exceeds kT are negligible, and therefore the collision frequency is seen to be proportional to T^3.

This does not, however, mean that the resistance will be proportional to T^3, since small f also means a very small deflexion of the electron. We have seen in (6.21) that the effectiveness of small-angle collisions is proportional to the square of the angle, and since in the present problem the angle is proportional to f, hence to kT, we expect the resistance to vary as T^5.

A resistance formula in agreement with this asymptotic law was derived by Bloch, assuming the electron motion to be isotropic, using the elastic waves for an isotropic continuum with a maximum frequency as in the Debye model, and assuming the phonons always to be in equilibrium. Bloch also assumes the constant C of (6.61) to be constant for longitudinal elastic waves and zero for transverse waves. The result relates the coefficient of the T^5 law to that of the linear resistance law at high temperatures and the characteristic Debye temperature Θ. This relation seems in better agreement with experiment than, with the many approximations contained in it, one would expect.

This is brought out particularly by the fact that the collision time for phonon-phonon collisions increases very rapidly at low temperatures. The thermal conductivity of non-metallic crystals increases at least as fast as the $1/T$ law which is valid at high temperatures; since the specific heat decreases as T^3, this means that the collision time varies at least as T^{-4}. On the other hand, the ratio of the number of phonons to the number of electrons in the border region is proportional to T^2. We have seen that the frequency of phonon-electron collisions per electron was proportional to T^3, and the frequency of the same collisions per phonon is therefore proportional to T. (No factor arises here from the smallness of the deflexion; while the collision deflects the electron only slightly, it completely destroys or produces the phonon.)

An important point is that, as we saw in § 2.4, collisions in which the total wave vector of all the phonons changes (Umklapp collisions) are very rare at low temperatures. Unless the electron-phonon interaction leads to such Umklapp processes, the equilibrium is therefore dependent on the former, and we would expect the resistance to decrease exponentially at low temperatures.

It is therefore very important to consider Umklapp collisions, i.e. those with $\mathbf{K} \neq 0$ in (6.46). Since at low temperatures f must be small for all collisions, we see that $\mathbf{k} - \mathbf{k}'$ must nearly equal $\mathbf{K}$, i.e. the initial and final states of the electron must be nearly equivalent. This means therefore that $\mathbf{k}$ must lie near a zone boundary, and $\mathbf{k}'$ near the opposite face of the zone.

In the particular case of a band with its minimum energy near $\mathbf{k} = 0$, and containing only a small number of electrons, states near the zone boundary are practically empty, the occupation number being of the order $e^{-D/kT}$, where D is again an energy of the order of the band width. This exponential is quite negligible at room temperatures, and in such a case the rate of Umklapp processes, and hence the resistance, is limited by the phonon-phonon collisions and should vary exponentially with temperature.

The same is true for a few empty places in a nearly full zone, except that in that case we must change, in accordance with Fig. 14 (a), the convention about wave vectors and with it the definition of Umklapp processes.

We can visualize this situation by saying that, in the first case, the collisions between electrons and phonons cause a drift of the phonons in the same direction as the electrons. Once the drift velocities are equal, further collisions are as likely to accelerate as to decelerate the electrons.

The same is true for a nearly full band, except that the current is now carried by a few holes in the electron distribution which, like positive carriers, tend to set up a phonon drift in the opposite direction to that in the other case.

In a metal in which there is one nearly empty and one nearly full band, the electrons in the one tend to produce a phonon drift in one direction, the holes in the other one in the opposite direction. It is usually stated that in this case equilibrium can always be established, but in fact this is true only if the number of electrons in the band equals that of the holes in the other.

It is likely that equilibrium can always be established if the energy surface $E(\mathbf{k}) = \eta$ intersects the zone boundary, since in that case a displacement of the whole electron distribution can be restored using up phonons of all directions.

It seems, however, unlikely that this is the case in the alkali metals, since these have body-centred cubic structures with one electron per atom. The zone is therefore just half full, and even on the simple approximation of § 4.2 (cf. Fig. 8) that energy surface which divides the zone in half is a cube which touches the boundary only with its corners. The real shape of the surface is likely to be more nearly spherical, and therefore clear of the boundary.

In such cases some other mechanism must be involved, and this is probably provided by the collisions between electrons. We have previously found these negligible at high temperatures; but since they give a collision time proportional to T^{-2} (cf. (6.7)) as compared to T^{-5} for the collisions with phonons, they must ultimately become dominant.

The condition for the existence of Umklapp collisions between electrons, i.e. collisions with $\mathbf{K} \neq 0$ in (6.64), is that the Fermi surface should extend at least half-way from the centre to the nearest zone boundary. For example, if the point half-way to the boundary (which is $\mathbf{K}/4$ if $\mathbf{K}$ is the reciprocal lattice vector associated with that part of the boundary) lies on the energy surface, we can take the transition

$$\mathbf{k}_1 = \mathbf{k}_2 = -\mathbf{k}'_1 = -\mathbf{k}'_2 = \tfrac{1}{4}\mathbf{K},$$

which satisfies (6.64) as well as the energy equation. For a Fermi surface of larger size such cases are more easily arranged by including a small angle between $\mathbf{k}_1$ and $\mathbf{k}_2$.

This condition is certainly satisfied in the alkalis, and it is therefore easy to understand why their conductivity does not increase exponentially at low temperatures. It is, however, very hard to see why, in these circumstances, the T^5 law, which took only phonon-electron collisions into account, and even the factor resulting from the simple Bloch theory, should be in such good agreement with observation. It is also hard to see why, amongst more complex structures, there should not occasionally be a case where the effective number of conduction electrons is so small that even the collisions between electrons are insufficient to cause a reasonable resistance.

We end this section with a remark about the thermal conductivity.

For high temperatures this problem is already solved by the remark that the collisions are in fact elastic, so that, according to § 6.2, the law of Wiedemann and Franz applies.

At very low temperatures, we can obtain the temperature variation in a similar way to the arguments in the case of electric conductivity. However, whereas the low-temperature electron-phonon collisions are inefficient in stopping an electron drift, because they deflect electrons only through small angles, they are quite efficient in adapting electrons to the temperature of their surroundings, since in each collision the electron energy may change by an amount of the order of kT. This is all that is required to change the temperature of the electron gas. Therefore the effective collision time for this purpose is proportional to T^{-3}, and not to T^{-5}, as before.

Since the specific heat of the electron gas is proportional to T, the thermal conductivity comes out proportional to T^{-2} at low temperatures.

This law is not subject to any doubt about the existence of Umklapp processes. The reason for this is that the definition of thermal conductivity refers to a situation where there is no electric current, and, as in § 6.2, we have to assume the presence of an electric field which will ensure this. There is therefore no electron drift, and no tendency to set up a phonon drift.

The phonons contribute, in principle, to the heat transport, but we have seen that they collide much more frequently with electrons than with each other, and therefore their collision time is much shorter than it would be in an insulator. Since their specific heat is small, the amount they contribute to the total heat transport is quite negligible.

6.8. Validity of assumptions

I stressed in § 6.3 the importance of the two assumptions which were used in our basic arguments. Assumption II in particular was stated in the form of the inequality $\hbar/\tau < kT$. But at high temperatures our estimate (6.74) showed that these quantities are of the same order of magnitude. Taking more precise figures for τ from measured conductivities, Mott and Jones (1936) quote the following collision times for the alkalis:

Li	Na	K	Rb	Cs	
0·86	3·1	4·4	2·7	2·1	$\times 10^{-14}$

which are to be compared with $\hbar/kT \sim 2\cdot 5 . 10^{-14}$. In general, Assumption II is therefore not well satisfied, and the position gets worse for impure specimens or alloys. Since $1/\tau$ is, at high temperatures, proportional to T, nothing is gained by changing the temperature in this region. However, at low temperatures, where the collision time increases more rapidly, the assumption is, in general, valid. For an imperfect specimen it may break down again at low temperatures, where the resistance is mainly due to impurities and tends to a constant limit.

It might therefore appear that all the theory of the transport phenomena stood on very shaky foundations. However, Landau has given an argument to show that, as long as one is dealing with elastic collisions only, one can obtain the results of the theory by a slightly different approach without using Assumption II.

In fact in all those cases in which the assumption is in doubt, namely at high temperatures and in conditions when the impurity resistance is dominant, we are dealing with elastic collisions, and they represent therefore the most critical case.

The argument by Landau (summarized by Peierls (1934 a)) starts with the consideration of electrons in a static field. This is assumed to contain the periodic potential of the perfect lattice as well as the effect of irregular disturbances. In this combined field, energy is still conserved, and we can define the number of states per unit energy, which we shall again denote by dZ/dE. We consider a large number of electrons distributed over these states so that the number of electrons in an interval dE is

$$2\frac{dZ}{dE} f(E)\, dE,$$

where $f(E)$ is an arbitrary function, not necessarily related to the Fermi function. Since we are not allowing for any processes which may change the energy, any such distribution will remain stationary.

Now imagine the same system in an infinitesimal electric field. We then conclude from the general principles of statistical mechanics that, provided the potential field contains a sufficiently irregular part, this will lead to a stationary state with a current proportional to the field. Moreover, since the electric field is capable of changing the electron energy by only infinitesimal amounts, the contribution to the total current from electrons of a particular energy will depend only on the distribution function $f(E)$ for that and immediately adjacent values of the energy.

We also know that the current must vanish if $f(E)$ is a constant, since a uniform distribution of electrons over all states in an energy band must behave like an insulator. Therefore the expression for the total current must be of the form

$$J = e^2 F \int Q(E) \frac{df}{dE}\, dE, \qquad (6.78)$$

where $Q(E)$ is some function of the energy. This is of the same form as (6.14), and, in fact, most of the theory of § 6.1 can be reproduced in this manner.

The main point is now this: from the definition (6.78), $Q(E)$ is related only to the properties of electrons with energies very close to E, and, since this is to be valid for any distribution function $f(E)$, the concept of temperature has not yet arisen. We now ask for the dependence of $Q(E)$ on the strength of the irregular part of the potential. If this is very weak, undoubtedly our Assumption II will be satisfied, and we can accept the results of the preceding sections, so that then

$$Q(E) = 2\overline{v_x^2}\tau \frac{dZ}{dE},$$

with the collision time τ inversely proportional to the square of the irregular potential and calculated according to § 6.3. That derivation is valid as long as $\hbar/\tau < kT$. But since the temperature appears neither in the mechanism nor in the definition of $Q(E)$, this inequality cannot represent the real limit of the proportionality between $1/\tau$ and the square of the potential. In fact, the only quantity of the dimension of an energy which occurs is the electron energy measured from the nearest band limit.

It follows from this that the real limit of the proportionality is given by the inequality

$$\frac{\hbar}{\tau} < \eta, \tag{6.79}$$

where η is measured from the nearest band limit. This is very much weaker than (6.41) and is practically always satisfied in metals.

It follows also that the work of § 6.6 is limited in validity only by (6.79), since in the high-temperature region the atomic displacements may be regarded as static (since the energy transmitted to the electrons is negligible) so that Landau's reasoning applies.

Some work is now in progress to illustrate this situation further. One result, contained in a recent paper by van Wieringen (1954), starts from the perturbation theory of § 6.3, but proceeds to include terms of fourth order, instead of only second order, in the potential W. It is found that no correction terms of the order $\hbar/\tau kT$ turn up, and the only corrections are of order $\hbar/\tau\eta$. While this, in itself, does not prove that the first approximation is adequate up to the limit (6.79), it provides a useful confirmation of Landau's reasoning.

It would seem, therefore, that in the ordinary theory of conduction the question of the limitation of the usual approach is a rather academic one. It looks as if the usual methods can be applied with correct results to practically all cases in which we can separate the forces into those determining the 'unperturbed' motion of the electron and the 'collisions', the effect of the latter being limited by a condition of the type (6.79). There are, however, cases in which such a division is not possible, for example in the resistance of alloys, and of liquids, and I shall later refer to such a case in the magneto-resistance problem.

VII

MAGNETIC PROPERTIES OF METALS

7.1. Paramagnetism

In this and the next few sections we shall be concerned with the effect of magnetic fields on metals in thermal equilibrium, and only later with their effect on transport phenomena.

The simplest problem is that of paramagnetism, i.e. the orientation of the electron spins by an external magnetic field. In a field of strength H, the energy of an electron is

$$E_l(\mathbf{k}) \pm \mu H, \tag{7.1}$$

where $\mu = e\hbar/2mc$ is the Bohr magneton, and the sign depends on whether the component of the electron spin in the field direction is $\frac{1}{2}$ or $-\frac{1}{2}$ units. The distribution of electrons over the states with energy (7.1) is now given again by the Fermi function, with a common value of η for electrons of both spin directions.

Since the Fermi function depends only on the difference $E - \eta$, this is equivalent to increasing the parameter η by μH for the electrons of one spin direction, and decreasing it for the other. If we therefore denote by $N(\eta)$ the total number of electrons without magnetic field, which is given by (4.54), then the number of electrons of the two spin directions is

$$\left. \begin{array}{l} N_+ = \tfrac{1}{2} N(\eta + \mu H) \\ N_- = \tfrac{1}{2} N(\eta - \mu H) \end{array} \right\} \tag{7.2}$$

In practice, the field is always so weak that we may approximate (7.2) by

$$\left. \begin{array}{l} N_+ = \tfrac{1}{2} N(\eta) + \tfrac{1}{2} \mu H \dfrac{dN}{d\eta} \\ N_- = \tfrac{1}{2} N(\eta) - \tfrac{1}{2} \mu H \dfrac{dN}{d\eta} \end{array} \right\} \tag{7.3}$$

Adding these equations, we see that the total number of electrons is $N(\eta)$, and since this must be the same number as for $H = 0$, it follows that to this order of approximation η is independent of H. (This could have been seen directly from the fact that η cannot depend on the sign of H and therefore cannot contain a first-order correction.)

If the temperature is low enough to make the electron gas degenerate, we may replace $N(\eta)$ by the leading term of (4.56), so that

$$N_+ = Z(\eta)+\mu H\left(\frac{dZ}{dE}\right)_\eta$$
$$N_- = Z(\eta)-\mu H\left(\frac{dZ}{dE}\right)_\eta \qquad (7.4)$$

The magnetic moment is therefore

$$M = \mu(N_+-N_-) = 2\mu^2 H\left(\frac{dZ}{dE}\right)_\eta. \qquad (7.5)$$

The paramagnetic susceptibility is therefore independent of temperature, and related to the coefficient of the linear term in the specific heat (cf. (4.57)). However, since we have neglected the interaction between the electrons, which probably affects the paramagnetism more than the specific heat, as we shall see in the next chapter, this relation should not be taken too literally.

Like the electronic specific heat, the paramagnetism is largest for a narrow band, and this also ceases to be degenerate at a lower temperature, so that we should use (7.3) rather than (7.4). Mott (cf. Mott and Jones, 1936, pp. 189 ff.) has applied this to the discussion of the transition elements, like Pd and Pt, in which there is an incomplete inner shell. The electronic states corresponding to those inner shells would, by themselves, give rise to narrow bands, since the overlap of the atomic wave functions between neighbouring atoms is small, and the approximation of § 4.2 therefore reasonable. The problem is complicated by the presence of the outer electron states, but the high value of dZ/dE due to the inner shell remains.

The result (7.5), showing the possibility of a large temperature-independent paramagnetism, was first derived by Pauli (1926) in a paper which started the development of the quantum theory of metals.

7.2. Diamagnetism of free electrons (Landau)

In the preceding section, I have dealt with the effect of a magnetic field on the electron spin. We must now consider its effect on the electron orbit. This is a much more difficult, but also much more interesting, problem. We shall first solve the case of free electrons, and discuss its relation to real metals later.

It is important to avoid a very natural error, which for a time gave rise to the impression that there is some fundamental difficulty about this problem.

§ 7.2   MAGNETIC PROPERTIES OF METALS   145

In classical mechanics, the projection of the electron orbit on a plane at right angles to the magnetic field is a circle of radius

$$r = \frac{mcv}{eH} = \frac{v}{2\Omega}, \qquad (7.6)$$

where v is the velocity of the electron in the transverse plane, c that of light, and

$$\Omega = \frac{eH}{2mc} \qquad (7.7)$$

is the Larmor frequency. Since a single circular orbit has, on the time average, a magnetic moment of

$$\frac{erv}{2c} = \frac{mv^2}{2H},$$

one is tempted to regard this as the magnetic moment per electron. The fact that this expression is independent of the charge, and inversely proportional to H, makes it obvious that the answer must be wrong. What has gone wrong is that we have assumed we are dealing with a number of complete circular orbits. In any finite volume, however, the electrons near the boundary cannot complete the circle. In addition to a number of complete circles, we must then consider also some circular arcs belonging to all those electrons whose orbits intersect the wall. These arcs amount together to a surface current which circles the volume in a sense opposite to that of the individual electron orbit, and which can easily be shown to cancel the effect of the complete orbits.

This argument might then give the impression that the answer to our problem is sensitive to the shape and nature of the surface. This, however, can be avoided if, instead of the moment due to each particle, we calculate the free energy F of the system, and then use the thermodynamic relation

$$M = -\frac{\partial F}{\partial H}. \qquad (7.8)$$

In classical mechanics, and using classical statistics,

$$F = -NkT \log \iint d^3p\, d^3v\, e^{-E(\mathbf{p},\mathbf{r})/kT}, \qquad (7.9)$$

where $E(\mathbf{p}, \mathbf{r})$ is the Hamiltonian function, expressing the energy of a particle as a function of momentum and coordinates. In a magnetic field

$$E(\mathbf{p},\mathbf{r}) = \frac{1}{2m}\left(\mathbf{p} - \frac{e}{c}\mathbf{A}\right)^2 + V(\mathbf{r}). \qquad (7.10)$$

If we substitute as integration variable

$$\mathbf{\Pi} = \mathbf{p} - \frac{e}{c}\mathbf{A} \qquad (7.11)$$

in place of **p** (**Π** is actually $m\mathbf{v}$, where **v** is the electron velocity) and perform the integration over **Π** first, it is at once evident that the vector potential **A**, and with it the magnetic field, has disappeared from the equation. The free energy therefore does not depend on H, and by (7.8) there is no magnetic moment. The same argument can be used if we employ Fermi statistics for the classical energy function, and it is therefore clear that any magnetic moment in thermal equilibrium must be due to the quantum effects in the motion of the particles.

We require, therefore, the energy levels of an electron in a magnetic field. It turns out to be convenient to work with the vector potential

$$A_x = 0, \qquad A_y = Hx, \qquad A_z = 0, \qquad (7.12)$$

which gives a uniform field of strength H in the z-direction. The Schrödinger equation is then

$$\frac{\partial^2 \psi}{\partial x^2} + \left(\frac{\partial}{\partial y} - \frac{ieHx}{\hbar c}\right)^2 \psi + \frac{\partial^2 \psi}{\partial z^2} + \frac{2mE}{\hbar^2}\psi = 0. \qquad (7.13)$$

Since this equation contains neither y nor z explicitly, we may assume

$$\psi(x,y,z) = e^{i(k_y y + k_z z)} u(x) \qquad (7.14)$$

and obtain for the function u the equation

$$u'' + \left\{\frac{2mE_1}{\hbar^2} - \left(k_y - \frac{eH}{\hbar c}x\right)^2\right\} u = 0, \qquad (7.15)$$

where

$$E_1 = E - \frac{\hbar^2}{2m} k_z^2 \qquad (7.16)$$

is the energy of the motion in the transverse plane.

(7.15) is the wave equation of an harmonic oscillator with frequency (cf. (7.7))

$$\frac{eH}{mc} = 2\Omega \qquad (7.17)$$

and centre

$$x_0 = \frac{\hbar c}{eH} k_y. \qquad (7.18)$$

Its energy eigenvalues are therefore

$$E_1 = (2n+1)\hbar\Omega = (2n+1)\mu H. \qquad (7.19)$$

This is thus independent of k_y. To find the number of states belonging to each eigenvalue, we have to specify boundary conditions. In the y- and z-directions we shall assume the cyclic condition, as before, which allows values for k_y and k_z which are multiples of $2\pi/L_2$ and of $2\pi/L_3$ respectively. It is not convenient to enforce a cyclic condition in the x-direction, since (7.13) and (7.15) depend on x and therefore have no periodic solutions. We assume, instead, that the metal is bounded by two walls, a distance L_1 apart, in the x-direction. If, then, L_1 is large compared to the extension of the oscillator function $u_n(x-x_0)$, which is of the order of the orbital radius (7.6), the functions for which the centre x_0 lies well inside the volume will not be affected by the presence of the walls. No solutions exist if x_0 lies well outside the volume; for a small range of x_0 near the wall, the presence of the wall modifies the oscillator eigenfunction and the energy value (7.19). Since in a field of 1,000 gauss, and an electron energy of 2 eV, the radius is about $5 \cdot 10^{-3}$ cm., this border region may usually be neglected.

Then the permissible interval of k_y is $(eH/\hbar c)L_1$. Hence we may write altogether

$$E(n, k_z) = (2n+1)\mu H + \frac{\hbar^2}{2m} k_z^2, \tag{7.20}$$

there being
$$\frac{L_1 L_2 eH}{2\pi \hbar c} \tag{7.21}$$

states for each energy. One easily verifies that the number of states in an interval large compared to $2\mu H$ is the same as for free electrons, so that the only new feature is their spacing.

It follows that, if we regard k_z as practically continuous, the number of states of energy below E is (apart from spin)

$$Z(E) = \frac{2(2m)^{\frac{1}{2}} L_1 L_2 L_3 eH}{(2\pi\hbar)^2 c} \sum_n \{E - (2n+1)\mu H\}^{\frac{1}{2}}, \tag{7.22}$$

where the sum is understood to extend over all those non-negative integers n for which the radicand is positive.

The free energy is (cf. (4.46))

$$F = N\eta - 2kT \int \frac{dZ}{dE} \log(1 + e^{(\eta-E)/kT}) \, dE, \tag{7.23}$$

or, on integrating by parts,

$$F = N\eta - 2 \int Z(E) f(E) \, dE, \tag{7.24}$$

where $f(E)$ is the Fermi function. Substituting from (7.22) we find, after a further integration by parts,

$$F = N\eta - A \int_{-\infty}^{\infty} \phi(\epsilon) \frac{d}{d\epsilon}\left(\frac{1}{1+e^{(\epsilon-\epsilon_0)/\theta}}\right) d\epsilon \tag{7.25}$$

with
$$\phi(\epsilon) = \sum_n (\epsilon - n - \tfrac{1}{2})^{\frac{3}{2}}, \tag{7.26}$$

the sum being again limited to those integers n for which the bracket is positive, and

$$\left. \begin{array}{c} \epsilon_0 = \eta/2\mu H, \qquad \theta = kT/2\mu H \\ A = \dfrac{16 m^{\frac{3}{2}}(\mu H)^{\frac{5}{2}} L_1 L_2 L_3}{3\pi^2 \hbar^3} \end{array} \right\}. \qquad (7.27)$$

We evaluate ϕ by means of Poisson's summation formula,† and find

$$\phi(\epsilon) = \sum_{l=-\infty}^{\infty} (-1)^l \int_0^{\epsilon} (\epsilon-x)^{\frac{3}{2}} e^{2\pi i l x}\, dx. \qquad (7.28)$$

For an evaluation of $\phi(\epsilon)$ this sum would converge rather slowly. However, the further integration required in (7.25) reduces the higher terms, provided θ is not too small. (7.25) shows also that ϕ is required only in the neighbourhood of ϵ_0, which is a large number provided

$$2\mu H \ll \eta. \qquad (7.29)$$

For $l \neq 0$, the integral in (7.28) can be reduced to

$$-\frac{3}{8\pi^2 l^2} e^{\pi i l \epsilon} \int_0^{\sqrt{\epsilon}} dy\, e^{-2\pi i l y^2} + \frac{3\epsilon^{\frac{1}{2}}}{8\pi^2 l^2} - \frac{\epsilon^{\frac{3}{2}}}{2\pi i l}. \qquad (7.30)$$

The first term is a rapidly oscillating function of ϵ which, for moderate temperatures,
$$kT \gg 2\mu H, \qquad (7.31)$$

gives a negligible contribution to (7.25). We therefore omit this term for the present. We then obtain

$$F = N\eta + A \int \left(\tfrac{2}{5}\epsilon^{\frac{5}{2}} - \frac{\sqrt{\epsilon}}{16} \right) \frac{d}{d\epsilon}\left(\frac{1}{1+e^{(\epsilon-\epsilon_0)/\theta}} \right) d\epsilon. \qquad (7.32)$$

The last factor has, as always, a steep maximum near ϵ_0 and, provided $\epsilon_0 \gg \theta$, i.e.
$$\eta \gg kT, \qquad (7.33)$$

we may take the value of the other factor at ϵ_0 outside the integral. This leaves

$$F = N\eta - \tfrac{2}{5} A \epsilon_0^{\frac{5}{2}} + \frac{A}{16} \epsilon_0^{\frac{1}{2}}. \qquad (7.34)$$

Remembering the definitions (7.27), we see that the first two terms are independent of H, and they should therefore represent the free energy in the absence of the field. The third term gives

$$F_1 = \frac{m^{\frac{3}{2}}(\mu H)^2 L_1 L_2 L_3 \sqrt{\eta}}{3\pi^2 \hbar^3 \sqrt{2}}, \qquad (7.35)$$

† Cf. Titchmarsh (1937), § 2.8.

§ 7.2   MAGNETIC PROPERTIES OF METALS   149

and, according to (7.8), a susceptibility

$$\chi_1 = \frac{M_1}{H} = -\frac{\sqrt{2}m^{\frac{1}{2}}\eta^{\frac{1}{2}}\mu^2}{3\pi^2\hbar^3} = -\frac{e^2}{12\pi^2\hbar c^2}\left(\frac{2\eta}{m}\right)^{\frac{1}{2}} \qquad (7.36)$$

per unit volume.

In terms of the wave number k_0 of the electrons in the border region, $\eta = (\hbar k_0)^2/2m$, so that

$$\chi_1 = -\frac{e^2 k_0}{12\pi^2 mc^2}. \qquad (7.37)$$

Strictly speaking, in the first two terms of (7.34), η should be allowed to depend on H, since in the equation

$$\frac{\partial F}{\partial \eta} = 0, \qquad (7.38)$$

which gives the correct defining equation for η, we should also include the field-dependent term. However, the change in η is then proportional to H^2, and since by (7.38) the variation of F with η is small, of the second order, this gives in F only a correction proportional to H^4, which is in general negligible.

The constant susceptibility which we have found can be seen to arise qualitatively from the fact that the energy levels (7.20) are bunched together. While their mean density is independent of H, the number of electrons that can be accommodated below any given level depends on whether this level coincides with one of the eigenvalues or falls between them. On the average we lose, since near $E = 0$ we always start with an empty interval. Hence on the average the energy of the electrons in the field is higher than without it.

It is easy to see that the diamagnetic moment M_1 is numerically equal to one-third of the paramagnetic moment (7.5) if the latter is also evaluated for free electrons. This relation is true not merely for high degeneracy of the Fermi gas, but also in the intermediate case and in the limit of Boltzmann statistics.

If we had required to know only the part of the moment which is proportional to H, somewhat simpler mathematical techniques would have sufficed. We shall now, however, evaluate the remaining term of (7.30) which is of interest at low temperatures. We therefore now assume that, instead of (7.31), $2\mu H$ is of the same order as kT. We retain, however, the assumption (7.29) so that we may take the integral in (7.30) to infinity. The first term then becomes

$$\frac{1-i}{4}\frac{3}{8\pi^2}\frac{1}{l^{\frac{1}{2}}}e^{2\pi i l \epsilon},$$

and this gives a further term in the free energy

$$F_2 = -A \sum_{l=1}^{\infty} (-1)^l \frac{3}{8\sqrt{2}\pi^2 l^{\frac{5}{2}}} \operatorname{re}\left[\int e^{i(2\pi l \epsilon - (\pi/4))} \times \frac{d}{d\epsilon}\left\{\frac{1}{1+e^{(\epsilon-\epsilon_0)/\theta}}\right\} d\epsilon \right],$$

(7.39)

where re means 'real part of'.

The integral is easily evaluated in the complex plane, by displacing the path of integration towards infinitely large positive imaginary values, and allowing for the poles at $(2n+1)\pi i$. The result is

$$-e^{i(2\pi l \epsilon_0 - (\pi/4))} \frac{2\pi^2 l \theta}{\sinh 2\pi^2 l \theta}.$$

This gives, to the free energy per unit volume, the contribution

$$F_2 = \frac{\sqrt{2}m^{\frac{3}{2}}(\mu H)^{\frac{5}{2}}kT}{\pi^2 \hbar^3} \sum_{l=1}^{\infty} \frac{(-1)^l}{l^{\frac{5}{2}}} \frac{\cos\{(\pi l \eta/\mu H)-(\pi/4)\}}{\sinh(\pi^2 l kT/\mu H)}. \quad (7.40)$$

The oscillatory behaviour of this expression is evidently related to the fact that, as the field varies, the border region of the Fermi distribution will sometimes be near one of the discrete levels and at other times between two levels. The denominator becomes large rapidly as soon as $kT/2\mu H$ is appreciable. In calculating the magnetic moment by (7.8), it is now essential to allow for the variation of η with H, since F_2 is very sensitive to small changes in η.

It would appear that this anomalous part of the diamagnetism is very hard to observe, since to make the denominator of the first term even as small as sinh 3, one has to use 80,000 gauss at 2° K. In addition, with $\eta \sim 2$ eV, the argument of the first cos in (7.40) would be about 10^4 at that field, so that the field would have to be kept constant to rather better than one part in 10^4 to observe any effect. We shall see later that for electrons in a lattice the position may in some cases be much easier.

7.3. Effect of a periodic field

It is now of interest to see how the diamagnetism is changed, if, instead of free electrons, we are concerned with electrons in a periodic potential. In this case we must include a potential energy term in the Schrödinger equation (7.13). Now since the force due to the magnetic field is always small compared to the force exerted by the lattice potential $V(\mathbf{r})$, one might be tempted to start from the solution without magnetic field and treat the magnetic terms as a small perturbation. This would not, however, be permissible, as is evident from the fact that even in the periodic field the particle orbits will be closed, and hence the energy levels presumably discrete. No disturbance that changes a continuous spectrum into a discrete one can be treated as a small perturbation. Indeed, the fact that H occurs in (7.13) multiplied by x means that far from the origin this term is very large.

We can choose any arbitrary point, say the centre of the nth lattice cell, as origin, if we modify the vector potential (7.12) accordingly. We must then change the wave function $\psi(\mathbf{r})$ to

$$\psi(\mathbf{r})e^{(ieH/\hbar c)x_n y}, \tag{7.41}$$

where x_n means the x component of $\mathbf{a}_n$. The function (7.41) satisfies, near $\mathbf{a}_n$, an equation which differs very little from that without magnetic field and it will therefore in that neighbourhood be similar to a combination of eigenfunctions without field.

This is the spirit of the approximation that may be applied to this equation in a number of ways. The first step then consists in taking the function (7.41) in the nth cell as a combination of electron states all belonging to the same band. The distortion of the wave by the magnetic field will then bring in also other bands just as, in expanding the wave function of an atom in a magnetic field, it will be found to contain small contributions from different atomic states. These terms will be referred to as non-diagonal terms.

As long as non-diagonal terms are neglected, one can show that the energy levels of the electron are given by changing the argument of the function $E_l(\mathbf{k})$ to a vector $\mathbf{\varkappa}$ whose components are now operators, satisfying the commutation law

$$\varkappa_x \varkappa_y - \varkappa_y \varkappa_x = \frac{ie}{\hbar c} H_z \tag{7.42}$$

(and corresponding equations for the other components), while $E_l(\mathbf{k})$ is the energy function of the periodic field as used previously. This is an extension of the free-particle case, in which we may regard (7.13) as expressing the energy in terms of the wave vector, and substituting for the wave vector

$$\mathbf{\varkappa} = \operatorname{grad} - \frac{ie}{\hbar c} \mathbf{A}, \tag{7.43}$$

where $\mathbf{A}$ is the vector potential. The components of (7.43) satisfy (7.42).

The proof of this statement is somewhat too long to be set out here. It was given by Peierls (1933) for the particular case of tight binding (§ 4.2), and a more elegant and general proof is contained in a thesis by Harper (1954).

If in the normal energy function of the conduction band we regard the wave vector components as non-commuting, we are left with an operator of which the exact eigenvalues are not easily found. It is, however, possible to calculate the high-temperature susceptibility (i.e. apart from any oscillating parts) without knowing the energy levels exactly. For this I must again refer to the original derivation by Peierls (1933) or a number of alternative forms, e.g. Wilson (1936). The result is that the constant part of the susceptibility is proportional to the integral over the border region of

$$R = \frac{\partial^2 E}{\partial k_x^2} \frac{\partial^2 E}{\partial k_y^2} - \left(\frac{\partial^2 E}{\partial k_x \partial k_y}\right)^2, \tag{7.44}$$

where the axis of the magnetic field has been taken as z-axis, and $E(\mathbf{k})$ is the energy function in the absence of a magnetic field. Since the volume of the border region is proportional to dZ/dE, the susceptibility is proportional to

$$\frac{dZ}{dE} \bar{R}. \tag{7.45}$$

For free electrons, this quantity is equal to

$$\frac{\hbar^2 k_0}{\pi^2 m}, \tag{7.46}$$

and for the general case the susceptibility is obtained by inserting (7.45) in place of (7.46) in (7.37). This may be carried out by defining an 'effective mass' from the equality of (7.45) and (7.46), choosing k_0 for this purpose as some average of the wave vector in the border region.

In particular, near the bottom of a symmetric zone, where the energy may be assumed proportional to k^2, and where we have already defined an effective mass in (4.44), we merely have to substitute m^* for m in (7.37). It is of interest that the diamagnetic effect increases for small effective mass, whereas the paramagnetic susceptibility (7.5) decreases in that case. Metals like bismuth, in which the total susceptibility is strongly diamagnetic, are therefore likely to have a small effective mass, or, from (7.44), an energy function whose gradient varies very rapidly with $\mathbf{k}$.

This explanation links up well with Jones's theory of the structure of Bi, § 5.4, since we have seen there that the structure arises from a near coincidence of the border of the Fermi distribution with a zone boundary created by the distortion of the structure from a simple cubic one. Now a glance at Fig. 13 shows that near a zone boundary created by a small distortion the second derivative of the energy with respect to the wave vector, and hence the effective mass, is particularly large. The same applies to certain types of alloy structure mentioned in § 5.4, which are also diamagnetic.

According to Jones, there is at least one band which is almost completely filled, containing only a few 'holes', and only a few electrons are overflowing into the next band. We are therefore here concerned with the neighbourhood of the energy maximum and minimum, respectively, and it seems reasonable to regard the energy as a quadratic function of **k**. At first sight, one would again be inclined to take a nearly isotropic function, since the Bi structure differs little from cubic symmetry. However, the energy maximum is likely to lie on the corners of the polyhedron defining the zone boundary, and the minimum of the next band near the middle of the faces. There are therefore several points with the same energy, which do not differ by a reciprocal lattice vector and therefore correspond to different electron states, but which are transformed into each other by a rotation of the crystal to an equivalent position. In this situation, as I pointed out in § 4.5, the energy surfaces near each minimum need not have any particular symmetry.

In spite of this we do not expect the susceptibility to be strongly dependent on direction, since the different portions of the energy surface will give contributions to (7.44) which will combine to a fairly isotropic overall result.

Turning now to the case of low temperatures, where the oscillating parts of the type (7.40) might be of importance, we require a knowledge of the actual grouping of the energy levels. For a general function of the non-commuting variables κ referred to in (7.42) this is not easy, but the problem is simple for the case of a quadratic function, i.e. near the maximum and minimum energy of a band. If this quadratic function is, in fact, isotropic, we can take over the whole theory of the preceding section, except for the value of the mass, which has to be replaced by the effective mass; this substitution has to be made also in μ where it occurs in (7.40), which is then replaced by

$$\mu^* = \frac{e\hbar}{2m^*c}.$$

For small effective mass the oscillating term will not only come in at higher temperatures, since the denominator of (7.40) is reduced, but will oscillate less rapidly, and be therefore more easily detectable.

If the energy is quadratic, but no longer isotropic, so that the energy surfaces are ellipsoids rather than spheres, one can reduce the eigenvalue problem to that solved previously, by a transformation which changes the scale of the different components of **k** by different factors. One may then show that the energy values near each minimum are still given by (7.20) provided μ, H, and m are changed by factors which depend on the direction of the magnetic field (Blackman (1938)). One therefore obtains a combination of terms like (7.40), containing different scale factors, one for each of the energy minima or maxima. Since now a change in orientation of the crystal will shift the different oscillatory turns relatively to each other in frequency, as well as altering their amplitudes, the overall result is far from isotropy.

Such an oscillatory behaviour of the susceptibility of bismuth had indeed been discovered by de Haas and van Alphen (1930) at low temperatures and in strong fields, and in fact this discovery led to the development of the theory of this and the preceding section. Further work by Shoenberg (1939) showed that the results, which are of fascinating complexity, can be used to provide a good deal of detailed information about the nature of the energy function near the maximum or minimum, and the number of electrons in each group. Later work (Shoenberg, 1952) has shown the existence of similar effects in other metals.

This raises the question whether one should also expect a de Haas–van Alphen effect if the conduction band is not nearly empty or nearly full. The characteristic features of the energy levels (7.20), (7.21) which we have used were that they were discrete (except for the motion in the field direction), equidistant, and of a very high multiplicity (7.21) proportional to H. Unpublished work by Harper has shown that going from the top or bottom of a band towards its centre, the energy levels cease to be equidistant; at the same time the sharp and degenerate levels are drawn out into narrow bands, whose width increases until, somewhere in the middle of the conduction band, they almost join together. The variability of the spacing will not affect the de Haas–van Alphen effect qualitatively, and the broadening will not affect it until the width becomes comparable with their spacing or with kT. Further work has to be done to decide whether there is a region in which the oscillatory behaviour becomes completely negligible. It would seem, at any rate, that, for sufficiently low temperatures, and with a very accurately uniform field, the effect should be found in most metals.

I stressed at the beginning that the non-diagonal terms have been omitted in this discussion. These must, in the limit of very strong binding, contain the susceptibility of the single atom as far as it is due to the electron orbits. They contain, however, also a cross term depending both on the function $E(\mathbf{k})$ and on transitions to different bands, which has no analogue either in the single atom or for free electrons. It has generally been assumed that these non-diagonal terms would contribute a constant susceptibility of the order of magnitude of that of single atoms, and would therefore be negligible both in cases of exceptionally high diamagnetism, as in Bi, and in the discussion of the de Haas–van Alphen effect.

In a recent paper Adams (1953) has, however, pointed out that when a small effective mass is due to a small gap between two energy bands as in Jones's theory of Bi, the non-diagonal terms may also be exceptionally large because the transitions to other bands are then associated only with small energy changes. It is not yet possible to estimate the total contribution from these terms, or to say whether they may also affect the oscillatory parts.

A further apparent difficulty is that the discrete nature of the energy levels, which we have used in the calculations of § 7.2, is connected with the fact that the classical electron orbits are periodic. Now if the electrons are subject to collisions, their orbits will not be periodic unless the collision time is longer. This might throw suspicion on all our results except when $\tau \gg 1/\Omega$, i.e.

$$\frac{\hbar}{\tau} \ll \mu H. \tag{7.47}$$

However, it was shown by Peierls (1933) that the collisions do not affect the constant susceptibility provided

$$\frac{\hbar}{\tau} \ll kT. \tag{7.48}$$

This condition is, in general, weaker than (7.47) and is satisfied, except at high temperatures, cf. § 6.8. It is, in fact, almost certain that (7.48) is not a necessary condition for the validity of the formula for the constant susceptibility, but that it is sufficient to have

$$\frac{\hbar}{\tau} \ll \eta. \tag{7.49}$$

This does not seem to have been proved. If true, it would make the collisions unimportant at all temperatures, except possibly for very poor conductors with a small number of conduction electrons.

For the de Haas–van Alphen effect, condition (7.48) is probably essential, since the level broadening due to the collisions is equivalent to an increase in temperature, as pointed out by Dingle (1952).

7.4. Hall effect and magneto-resistance

In discussing the effect of a magnetic field on transport phenomena, we again omit non-diagonal terms. If we then introduce again the non-commuting components of $\varkappa$, it follows from the commutator (7.42) that

$$\left.\begin{aligned}\frac{d\kappa_x}{dt} &= -\frac{i}{\hbar}(E\kappa_x - \kappa_x E) = -\frac{eH}{\hbar^2 c}\frac{\partial E}{\partial \kappa_y} = -\frac{eH}{\hbar c}v_y \\ \frac{d\kappa_y}{dt} &= -\frac{i}{\hbar}(E\kappa_y - \kappa_y E) = \frac{eH}{\hbar^2 c}\frac{\partial E}{\partial \kappa_x} = \frac{eH}{\hbar c}v_x\end{aligned}\right\} \quad (7.50)$$

if the magnetic field is in the z-direction. Here **v** is again the function defined by (4.41) with $\varkappa$ as argument. Now we may describe the electrons by wave packets in which the values of both κ_x and κ_y are approximately fixed, and this will be adequate for a statistical description provided the extent in **k** space of these wave packets is small compared to the border region, which amounts in order of magnitude to the condition

$$\mu H \ll kT. \quad (7.51)$$

Subject to this condition we may from now on replace $\varkappa$ by its mean over the wave packet, whose components commute, and revert to the notation **k**. (7.50) then confirms the equation for acceleration by a field, which we surmised in (4.43).

To obtain the effect of the field on the current, we must add in the Boltzmann equation besides (6.9) a term

$$\frac{eH}{\hbar c}\left(v_y \frac{\partial n}{\partial k_x} - v_x \frac{\partial n}{\partial k_y}\right). \quad (7.52)$$

We shall again start with the assumption of a definite collision time, which may still be a function of the energy. Then we have in place of (6.13), omitting again $\bar{n}_1$ for the same reason as before,

$$\frac{1}{\tau}n_1 + \frac{eH}{\hbar c}\left(v_y \frac{\partial n_1}{\partial k_x} - v_x \frac{\partial n_1}{\partial k_y}\right) = -e(F_x v_x + F_y v_y)\frac{df}{dE}. \quad (7.53)$$

For free electrons, this simplifies to

$$\frac{1}{\tau}n_1 + \frac{eH}{mc}\left(v_y \frac{\partial n_1}{\partial v_x} - v_x \frac{\partial n_1}{\partial v_y}\right) = -e(F_x v_x + F_y v_y)\frac{df}{dE}. \quad (7.54)$$

The bracket on the left-hand side represents an operation which gives zero when applied to any function of the energy, and which turns a linear function of the velocities into another such linear function. It follows therefore that the equation can be solved by putting

$$n_1 = g_x v_x + g_y v_y, \quad (7.55)$$

where g_x, g_y are functions of the energy. Inserting this and comparing the angular dependence of the terms on either side,

$$\left.\begin{aligned}\frac{1}{\tau}g_x - 2\Omega g_y &= -eF_x\frac{df}{dE} \\ \frac{1}{\tau}g_y + 2\Omega g_x &= -eF_y\frac{df}{dE}\end{aligned}\right\}, \qquad (7.56)$$

where Ω is the Larmor frequency, defined in (7.7). Hence

$$\left.\begin{aligned}g_x &= \frac{-e\tau(F_x + 2\Omega\tau F_y)}{1 + 4\Omega^2\tau^2}\frac{df}{dE} \\ g_y &= \frac{-e\tau(F_y - 2\Omega\tau F_x)}{1 + 4\Omega^2\tau^2}\frac{df}{dE}\end{aligned}\right\}. \qquad (7.57)$$

If we have current flowing along a wire, the direction of the current is given. If we take this as the x-direction, J_y must vanish. Now since (cf. (6.14))

$$\left.\begin{aligned}J_x &= 2e\int \overline{v_x^2}\frac{dZ}{dE}g_x\,dE \\ J_y &= 2e\int \overline{v_y^2}\frac{dZ}{dE}g_y\,dE\end{aligned}\right\}, \qquad (7.58)$$

and since the factor df/dE vanishes except near $E = \eta$, this means that the other factor in g_y must vanish at η,

$$F_y = 2\Omega F_x \tau(\eta). \qquad (7.59)$$

It is customary to express the transverse potential gradient in terms of the longitudinal current and the magnetic field, thus defining the Hall coefficient R:

$$R = \frac{F_y}{HJ_x} = \frac{e\tau(\eta)}{mc\sigma}, \qquad (7.60)$$

or, inserting for the conductivity σ from (6.16),

$$R = \frac{1}{ecN}. \qquad (7.61)$$

This well-known result would seem to indicate that the Hall coefficient is a measure of the number of conduction electrons per unit volume.

Under the same assumptions we see that, for energies near η, the first equation (7.56) is independent of the magnetic field. Since only these energies are of importance for (7.58), it therefore follows also that the magneto-resistance effect is negligible.

These results hold good for isotropic conditions, even if the scattering probability is a function of the scattering angle, as in (6.17). This follows because the distribution (7.56) is a spherical harmonic of first order, so that, as in the case of electric conductivity, the collision time τ_1 from (6.21) replaces τ in our equations.

The reason for the absence of magneto-resistance has been particularly simply expressed by Wilson (1939): In the isotropic case each electron will have the same mean velocity. The transverse electric field will adjust itself so as to cancel the transverse force due to the magnetic field of an electron of that velocity. Hence the electrons are subject only to the longitudinal field and to collisions.

If, however, the mean velocity is different for different groups of electrons, then the transverse force balances only on the average, and each electron is still travelling on a curved path. This can be illustrated simply if we assume there are electrons in two different bands, each isotropic, and each with its own collision time. We assume that the exchange of electrons between the bands is slow. Then for each band

$$F_x = \frac{1}{\sigma} J_x - RHJ_y,$$

$$F_y = \frac{1}{\sigma} J_y + RHJ_x,$$

or, solving for the current,

$$J_x = \frac{\sigma F_x + RH\sigma^2 F_y}{1 + R^2 H^2 \sigma^2}, \qquad J_y = \frac{\sigma F_y - RH\sigma^2 F_x}{1 + R^2 H^2 \sigma^2}. \qquad (7.62)$$

The resultant current is the sum of the currents in the two bands. Adding two expressions like (7.62), and again requiring $J_y = 0$, we find, after some tedious algebra,

$$F_y = \frac{R_1 \sigma_1^2 + R_2 \sigma_2^2 + R_1 R_2 \sigma_1^2 \sigma_2^2 H^2 (R_1 + R_2)}{\sigma_1 + \sigma_2 + \sigma_1 \sigma_2 (R_1^2 \sigma_1 + R_2^2 \sigma_2) H^2} HF_x, \qquad (7.63)$$

and for the relative increase in resistivity

$$\frac{\Delta \rho}{\rho_0} = \frac{\sigma_1 \sigma_2 (\sigma_1 R_1 - \sigma_2 R_2)^2 H^2}{(\sigma_1 + \sigma_2) + H^2 \sigma_1^2 \sigma_2^2 (R_1 + R_2)^2}. \qquad (7.64)$$

The first of these equations gives the Hall coefficient if we express F_x in terms of J_x. Usually the Hall effect is measured in moderately weak fields, and then we may neglect the terms in H^2 in (7.63), and also put $F_x = J_x/(\sigma_1 + \sigma_2)$. Then

$$R = \frac{R_1 \sigma_1^2 + R_2 \sigma_2^2}{(\sigma_1 + \sigma_2)^2}. \qquad (7.65)$$

As regards the change in resistance, we notice that it vanishes again when the product of Hall coefficient and conductivity is the same in the two bands. From (7.61) and (6.16) this means that the 'mobility'

$e\tau/m$ must be the same for both bands, in accordance with the qualitative argument given above.

Otherwise, there is an increase in resistance which is initially proportional to H^2 but saturates at fields for which $H\sigma R$ is larger than unity for both bands. This amounts dimensionally to the condition that the collision time for each band be larger than the Larmor frequency, or again to (7.47). While for the constant susceptibility the relative magnitude of collision time and Larmor period was unimportant, we see that in the magneto-resistance it plays an important part.

If we are dealing with a nearly full band, so that there are only a few places near an isotropic energy maximum, the theory goes through as before, but, as there are now positive 'holes' carrying the current, the sign of the Hall effect is reversed, as is indicated by the occurrence of the first power of the charge in (7.61).

If a metal contains some positive holes in one band and a few electrons in another, the Hall coefficient is given by (7.65) and it is clear that the sign now depends on the numbers and mobilities of electrons and holes.

All this discussion applies to the isotropic case. For a more general case we would have to return to equation (7.53). Its solution can no longer be given in closed form, though it can be given in the form of a power series in either ascending or descending powers of H, the limits of validity of either being given again by the ratio between collision time and Larmor period. The magneto-resistance again does not vanish, and we may interpret this as due to the fact that, although we have assumed the collision time constant, the effective mass is different for electrons moving in different directions. One can also show that for very high fields the resistance again tends to a constant limit.

However, there is little point in putting much effort into a discussion of this equation, since the assumption of a constant collision time is certainly not justified in the anisotropic case. We may no longer argue, as we did in the case of the Wiedemann–Franz law, that we are dealing with a deviation from equilibrium which varies over the Fermi zone as a component of the electron velocity, so that we may use the same average of the collision time as in the electric conductivity. This is true in the isotropic case, where we found a solution of the form (7.55), but it is easily seen by inserting (7.55) in the general anisotropic equation (7.53) that this does not give a solution.

One interesting application of the arguments of this section is to demonstrate the anisotropy of metals like the alkalis. Here it is almost certain that all electrons are in the same band, and if their energy surfaces and their collisions with lattice waves were isotropic, one should expect zero magneto-resistance. In fact, their magneto-resistance coefficients are rather smaller than those of other metals, but not in order of magnitude.

In this discussion we have neglected effects arising from the quantum character of the electron motion. At not too low temperatures this can be allowed for by a perturbation treatment. A calculation of this kind was carried out by Titeica (1935) who showed that there are effects of the order of $\mu H/\eta$, but not, as one might have expected, of the order of $\mu H/kT$. Titeica's method is valid only when the field is strong enough for the Larmor period to be less than the collision time. This has been extended to the general case in unpublished calculations by van Wieringen. The magneto-resistance due to this cause is always much less than the 'detour' effect considered before.

These quantum effects become particularly interesting at very low temperatures, where Schubnikow and de Haas (1930) found anomalies in the magneto-resistance of Bi in parallel with the de Haas–van Alphen effect. No theory of this case is as yet available, and it represents a most fascinating problem in quantum statistics, since the conditions for the validity of a Boltzmann equation, which we discussed in § 6.8, are probably not satisfied.

Note added January 1956

The work by Harper referred to at the bottom of page 151 has been published. *Proc. Phys. Soc.* A, **68**, 874, 1955 and **68**, 879, 1955.

VIII

FERROMAGNETISM

8.1. The Weiss model

THE term ferromagnetism is applied to a magnetic behaviour of which iron is a typical example and which also is found in cobalt and nickel, and in certain alloys and compounds.

It was recognized by Weiss that the properties of these substances can be understood if one assumes that they consist of small domains, each of which is usually magnetized. In an apparently non-magnetic piece of iron, the directions of magnetization of the domains are distributed at random, so that the total moment is zero. The process of magnetization consists in altering the directions of the magnetization in the domains, without altering the intensity of magnetization in each. It is therefore clear that the magnetization of a domain or 'intrinsic magnetization' at any temperature is one of the most fundamental properties of a ferromagnetic.

Weiss also realized that the existence of such an intrinsic magnetization must be due to a strong interaction of the elementary magnetic moments contributed by the different atoms. This interaction must have a tendency to make the moments take the same direction.

Such an interaction is, in principle, caused by magnetic forces, but these depend sensitively on the external shape of the body, and, above all, they are, as we shall see, much too weak to explain an intrinsic magnetization except at very low temperatures. Weiss therefore postulated the existence of a 'molecular field' of unknown origin which was proportional to the magnetization, and acted on each individual magnet like a strong magnetic field.

It is now known from measurements of the gyromagnetic effect, i.e. of the ratio of mechanical angular momentum to magnetic moment, that the elementary magnets are electron spins. For the present purpose I shall assume that we may regard each magnet as a single spin, which is therefore capable of two orientations. In this case (which, of course, is not precisely the picture used by Weiss) we may then say that the spin will have a different energy according to whether it is in the direction of prevalent magnetization, or opposite to it. The energy is

$$E = \pm \mu H_m, \tag{8.1}$$

where H_m is the Weiss molecular field. The distribution of spins over the two directions is then given by Boltzmann's factor, so that

$$\frac{N_+}{N_-} = e^{2\beta\mu H_m}. \tag{8.2}$$

The total moment is therefore

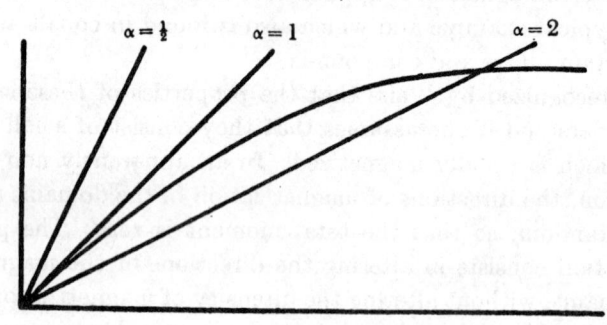

$$M = \mu(N_+ - N_-) = \mu N \frac{e^{\beta\mu H_m} - e^{-\beta\mu H_m}}{e^{\beta\mu H_m} + e^{-\beta\mu H_m}} = N\mu \tanh\frac{\mu H_m}{kT}. \tag{8.3}$$

FIG. 15.

We now assume that the molecular field is proportional to the magnetization, so that

$$H_m = \frac{C(N_+ - N_-)}{N}. \tag{8.4}$$

Combining (8.3) and (8.4) we have the equation

$$\frac{M}{M_0} = \tanh\alpha\frac{M}{M_0}, \tag{8.5}$$

where

$$M_0 = N\mu; \qquad \alpha = \frac{\mu C}{kT}. \tag{8.6}$$

Fig. 15 shows $y = x/\alpha$ and $y = \tanh x$ for several values of α.

An intersection represents a solution of (8.5), and the ordinate is M/M_0. It is evident that no intersection with $M \neq 0$ exists, if $\alpha \leqslant 1$, i.e. if

$$T \geqslant \frac{\mu C}{k} = \Theta. \tag{8.7}$$

This determines the 'Curie point' Θ, above which the spontaneous magnetization vanishes. Below Θ there always exists a solution, which for decreasing temperature rapidly approaches saturation. It is easily seen that, for temperatures just below Θ,

$$\left(\frac{M}{M_0}\right)^2 = 3\frac{\Theta - T}{\Theta}, \tag{8.8}$$

whereas at very low temperatures

$$\frac{M}{M_0} = 1 - 2e^{-\Theta/T}. \tag{8.9}$$

Equation (8.8) shows that the magnetization disappears continuously, but has an infinite temperature derivative at the Curie point. This sharply defined temperature is a characteristic feature of a 'co-operative' phenomenon, in which the simultaneous interaction of very many particles plays a part.

In quantitative detail (8.5), and the limiting laws (8.8) and (8.9), are not in good agreement with observation, but they do show the right qualitative behaviour. In order to account for the order of magnitude of the Curie point, which in iron is about $1,000°$, μC, the energy of one spin when all others are parallel, should be of the order of 0.1 eV. If the molecular field was a magnetic effect, this would therefore require a field of the order of $1.5 \cdot 10^7$ gauss, which is much too high.

The correct explanation was given by Heisenberg. He pointed out that an effect of just this kind would be provided by electron exchange effects. In any electronic system containing two electrons in different orbital states, we may have the electron spins either parallel or opposite. In the first case we have to form the antisymmetric, in the second the symmetric combination of the orbital wave functions. The two belong to different spatial distributions of the electrons and in the case of parallel spin the electrons are less likely to be very close to each other, with a corresponding reduction in their interaction energy. The same cause is responsible for the separation between different multiplet terms in atomic spectra. Of several multiplets with the same configuration (i.e. the same individual levels for the separate electrons) the one with the highest multiplicity, and therefore the largest resultant spin, in general has the lowest energy, and the energy differences are of the order of an eV and therefore sufficient for the effect here required.

The electron exchange provides therefore a mechanism that can explain the existence of ferromagnetism. This opens the way, in principle, to a quantitative treatment. In the next few sections we shall consider two methods that have been applied to this problem in the past, starting from different approximations. These are known, respectively, as the theory of spin waves, and the collective electron theory. Of these the latter follows more closely the lines of the approximation to a metal which we have used in previous chapters, but is the more difficult one mathematically, and little is known as yet about the properties of its solutions. It will therefore be more convenient to discuss the spin-wave theory first.

8.2. The spin-wave theory. One dimension

It is evident that ferromagnetism depends on the interaction between the electrons and that we therefore must not neglect the interaction between electrons, as we have mostly done up to now. We must therefore consider the whole system of ions and electrons, and one simple way of doing this is to start from neutral atoms. We imagine that N neutral atoms, each for simplicity containing one valency electron, are brought together to form a crystal.

If we may still regard the atoms as well separated, and each electron strongly bound to its atom, the situation is similar to that of the strong-binding approximation of § 4.2. However, in that case we took the electrons as moving independently, so that the chance of finding two electrons in the same atom would be no less than that of finding them in two different places. The electron interaction will reduce the likelihood of having two electrons located in the field of the same ion, and for simplicity we exaggerate this by assuming that it may never happen. In that case each atom contains just one electron, and the only freedom we have is in arranging their spins.

This model, which is the same as that used in the Heitler–London theory of the hydrogen molecule, cannot give any conductivity, since it assumes that the electrons cannot pass over each other, and the probability of a whole column of electrons making a simultaneous quantum jump is extremely small.

To employ such a model in the case of a metal is not as inconsistent as might appear at first sight. Mott and Jones (1936) have shown that in the ferromagnetic metals, in which the atomic $3d$ and $4s$ levels lie very close together, one must regard both these states as partly filled. If one considered only the $3d$ states, they would give rise to a very narrow band, since the overlap of neighbouring wave functions is small, whereas the $4s$ states by themselves give rise to a wide band. If we constructed Bloch wave functions by the method of § 4.2, suitably amended to include several different atomic states, we would find that the wave functions are in general mixtures of the d and s states, but there will be only a small region within the conduction band in which the $3d$ states will play a prominent part. It is therefore a reasonable approximation to distinguish $3d$ and $4s$ electrons even in the crystal, and Mott and Jones also show that it is mostly the $3d$ electrons which are responsible for the

§ 8.2 FERROMAGNETISM

magnetism, whereas the 4s electrons, which form a wider band, and therefore have greater velocities, would be expected to be responsible for the conductivity. The Heitler–London model would then be intended to describe only the 3d electrons.

To explain the procedure, we assume that there is only one electronic state per atom, and we consider the case of a linear chain of atoms.

If the atoms were well separated, the state of the system would have to be described by means of the atomic wave function $\phi_n(x)$ of the nth atom, in which only the coordinate x of the valency electron is shown explicitly, and a spin function χ_μ of that electron, where μ indicates some component of the spin and may equal $\frac{1}{2}$ or $-\frac{1}{2}$. We may now specify the state by writing down the spin component of each atom, i.e. by a set of numbers $\mu_1, \mu_2, ..., \mu_N$. For brevity we shall use the symbol $\{\mu\}$ for the whole set of numbers.

The wave function belonging to this set is then

$$\Phi\{\mu\} = \frac{1}{\sqrt{(N!)}} \begin{vmatrix} \phi_1(x_1)\chi_{\mu_1}(1) & \phi_2(x_1)\chi_{\mu_2}(1) & \cdots & \phi_N(x_1)\chi_{\mu_N}(1) \\ \vdots & \vdots & & \vdots \\ \phi_1(x_N)\chi_{\mu_1}(N) & \phi_2(x_N)\chi_{\mu_2}(N) & \cdots & \phi_N(x_1)\chi_{\mu_N}(N) \end{vmatrix} \quad (8.10)$$

where $x_1, x_2, ..., x_N$ are coordinates of the N electrons, $\chi_{\mu_1}(1)$ is the spin function of the first electron having component μ_1. This function is antisymmetric in the electrons as required by Pauli's principle, and the factor in front of the determinant secures the correct normalization.

If the atoms are now placed at a finite distance from each other, (8.10) is no longer an energy eigenfunction, since it ignores the interaction of each electron with the field of other atoms and the electrons in them. We therefore look for the best combination of functions of the class (8.10) which will approximate to an exact solution, and to do this we proceed as in § 4.2 to minimize the expectation value of the total energy.

We define a combination

$$\Psi = \sum_{\{\mu\}} A\{\mu\}\Phi\{\mu\} \quad (8.11)$$

and we determine the coefficients by minimizing the energy

$$E - E_0 = \frac{(\Psi^*(H-E_0)\Psi)}{(\Psi^*\Psi)} = \frac{\sum A^*\{\mu\}A\{\mu'\}(\Phi^*\{\mu\}(H-E)\Phi\{\mu'\})}{\sum A^*\{\mu\}A\{\mu'\}(\Phi^*\{\mu\}\Phi\{\mu'\})}, \quad (8.12)$$

where H is the Hamiltonian, and the brackets imply an integration over all electron coordinates and summation over spin variables. E_0 is the energy of N single atoms. This expression is stationary if the A satisfy the equation

$$\sum_{\{\mu'\}} [(E-E_0)(\Phi^*\{\mu\}\Phi\{\mu'\}) - (\Phi^*\{\mu\}(H-E_0)\Phi\{\mu'\})]A\{\mu'\} = 0. \quad (8.13)$$

Now in forming the product of two functions (8.10) we obtain a number of terms containing the product of two different atomic wave functions

$$\phi_n^*(x)\phi_{n'}(x). \quad (8.14)$$

For $n \neq n'$ these products are always small, since in the spirit of the approximation we take the atomic distance as rather larger than the atomic radius, so that the overlap between the wave functions is small. The largest terms will therefore be those in which we take corresponding terms from the two determinants, so that the electron which occurs in one factor in the nth atom is in the same atom in the other factor. Such terms, however, occur only if $\mu_{n'} = \mu_n$, since otherwise the spin functions are orthogonal. The largest contribution to (8.13) therefore comes from the term in which $\{\mu'\} = \{\mu\}$. In that case the large terms give

$$(\Phi^*\{\mu\}\Phi\{\mu\}) = 1, \quad (8.15)$$

$$(\Phi^*\{\mu\}(H-E_0)\Phi\{\mu\}) = \sum_{n \neq n'} \alpha_{n,n'}, \quad (8.16)$$

where

$$\alpha_{n,n'} = \iint |\phi_n(x)|^2 |\phi_{n'}(x')|^2 [U_{n'}(x) + \tfrac{1}{2}V(x,x') + \tfrac{1}{2}W]\,dxdx' \quad (8.17)$$

is the mean energy change of the nth electron due to the presence of the atom n'. $U_{n'}(x)$ is the potential of the core of the atom n', as in § 4.2, $V(x, x')$ is the interaction between the two electrons, and W that of the cores. The factor $\tfrac{1}{2}$ in the last two terms is there because otherwise the interaction between any two particles would be counted twice. The integral $\alpha_{n,n'}$ depends only on the distance between n and n' and decreases rapidly with that distance. We neglect all terms except those for adjacent atoms. If we write α for $\alpha_{n,n+1}$, then (8.16) is $2N\alpha$.

We now go one step further and include terms with overlap factors of the type (8.14), provided they do not contain more than two such factors, and we allow only overlap between adjacent atoms. In the diagonal term $\{\mu\} = \{\mu'\}$, we then get a further contribution by multiplying two products from the determinant (8.10) in which two adjacent electrons with the same spin have been interchanged. (If the spins are different, the product vanishes because of the orthogonality of the χ.) This contributes to (8.15) the term $-r|\beta|^2$, where

$$\beta = \int \phi_n^*(x)\phi_{n+1}(x)\,dx \quad (8.18)$$

§ 8.2 FERROMAGNETISM 167

and r is the number of pairs of adjacent parallel spins. The contribution to (8.16) is $-r\gamma$, where

$$\gamma = \iint \phi_n^*(x)\phi_{n+1}^*(x')[U_n(x')+U_{n+1}(x)+V(x,x')+W] \times \\ \times \phi_n(x')\phi_{n+1}(x)\,dxdx'. \quad (8.19)$$

In the bracket of the 'exchange integral' the first two terms are negative, whereas the other two are positive. The factors outside the bracket taken together are positive when $x = x'$, but, if the wave function has nodes, will be negative in other parts. We shall for the present take γ as positive, since this is the case in which we shall find ferromagnetic behaviour.

Terms of the order β and γ now appear also when $\{\mu\}$ and $\{\mu'\}$ are not identical, but differ merely by interchanging two opposite spins. In that case
$$\left.\begin{array}{l}(\Phi^*\{\mu\}\Phi\{\mu'\}) = -\beta^2 \\ (\Phi^*\{\mu\}(H-E_0)\Phi\{\mu'\}) = -\gamma\end{array}\right\}. \quad (8.20)$$

Combining all these results, we find that (8.13) takes the form

$$[(E-E_0)(1-r\beta^2)-2N\alpha+r\gamma]A\{\mu\} = \sum_{\{\mu'\}}[(E-E_0)\beta^2-\gamma]A\{\mu'\} \quad (8.21)$$

provided it is understood that the sum on the right includes only those spin arrangements which differ from $\{\mu\}$ by the exchange of two adjacent opposite spins.

It is now plausible that we are justified in neglecting β^2, which appears always multiplied by $(E-E_0)$, which is itself of the first order in the small quantities α, β^2, γ, since without the interactions and overlap evidently $E = E_0$. We also note that $N-r$ is the number of adjacent pairs of unlike spins, which is just the number of terms on the right-hand side.

With the abbreviation

$$\epsilon = E - E_0 - 2N(\alpha - \gamma/2) \quad (8.22)$$

the equation for the A is finally

$$\epsilon A\{\mu\} = \gamma \sum_{\{\mu'\}}(A\{\mu\}-A\{\mu'\}). \quad (8.23)$$

Here the sum is to be understood with the same restriction as before. It therefore, in particular, does not connect spin arrangements which have different values of the total component

$$m = \sum_n \mu_n. \quad (8.24)$$

(This is true quite generally and not merely to the approximation used here.) We may therefore consider each m value separately. The simplest case is $m = N$, when all spins are parallel. Then the sum in (8.23) is empty, and $\epsilon = 0$.

Next take $m = N-1$, i.e. one spin reversed. If its position is denoted by n, we may use n to identify the spin arrangement $\{\mu\}$, and (8.23) becomes
$$\epsilon A_n = \gamma(2A_n - A_{n+1} - A_{n-1}) \tag{8.25}$$
with the solution
$$A_n = \text{constant} \times e^{igan}, \qquad \epsilon(g) = 2\gamma(1 - \cos ga). \tag{8.26}$$
Here we have again included the atomic distance a in the definition of the phase factor for the convenience of later generalization. In analogy with §§ 1.5 and 4.2 it follows again that g must be real and a multiple of $2\pi/Na = 2\pi/L$, where L is the length of the chain, and that it will be convenient to assume it to lie between $-\pi/a$ and π/a. The energy values form a practically continuous band from $\epsilon = 0$ to $\epsilon = 4\gamma$.

This result shows up one of the important qualitative differences between the Weiss model and the present picture. Whereas in the Weiss model the energy required to reverse a spin from the completely saturated state is equal to the constant quantity $4k\Theta$, we now find a continuous distribution of energy values. This tends to increase the number of 'wrong' spins at low temperatures. In fact, as we shall see, in the one-dimensional case it is so easy to reverse a spin that one does not get any ferromagnetic behaviour at all.

Consider now the case $m = N-2$. Then we characterize the spin arrangement by stating the positions n_1, n_2 of the two spins which are 'wrong'. We therefore have an amplitude $A(n_1, n_2)$ which is defined only when $n_1 \neq n_2$ and we may also assume that $n_1 < n_2$. Then (8.23) takes the form
$$\epsilon A(n_1, n_2) = \gamma[4A(n_1, n_2) - A(n_1+1, n_2) - A(n_1-1, n_2) -$$
$$- A(n_1, n_2+1) - A(n_1, n_2-1)] \quad (n_2 \neq n_1+1), \tag{8.27a}$$
$$\epsilon A(n, n+1) = \gamma[2A(n, n+1) - A(n-1, n+1) - A(n, n+2)]. \tag{8.27b}$$

The right-hand side of (8.27 a) is a sum of two operators, of which the first acts only on n_1, the second only on n_2, and consequently we can find a solution of this equation in the form of a product
$$e^{ia(g_1 n_1 + g_2 n_2)}, \tag{8.28}$$
but this will not satisfy (8.27 b). We can, however, meet this second requirement by combining two solutions of this type with suitable coefficients. This situation is to be interpreted as follows. (8.28) describes two separate 'wrong spins' or 'spin waves' travelling along the chain. We may form an almost stationary state by building wave packets for

each of the two spins, and, if they are well separated, the amplitude $A(n_1, n_2)$ vanishes near $n_1 = n_2$, and therefore (8.27 b) does not arise. However, the two wave packets will travel with a velocity which will obviously be the group velocity

$$v(g) = \frac{1}{\hbar}\frac{\partial \epsilon}{\partial g} = \frac{2\gamma a}{\hbar}\sin ga \qquad (8.29)$$

for each. These will in general be different, and so the two spin waves will meet after some time. The correction required by (8.27 b) then results in scattering. In this scattering process the translational invariance of the problem will, in complete analogy with previous collision problems, lead to the equation

$$g_1' + g_2' - g_1 - g_2 = K, \qquad (8.30)$$

where K in this linear case is either 0 or $2\pi/a$.

However, solutions of the form (8.28) with real g_1, g_2 are not the only solutions of (8.27 a) which are physically reasonable. This is because by definition $n_2 - n_1$ is positive, and if we give to g_1 and g_2 equal and opposite imaginary parts, this may lead to a solution which decreases exponentially as $n_2 - n_1$ is large. This leads to a solution of the form

$$A(n_1, n_2) = \text{constant} \times e^{iGa(n_1+n_2)-ba(n_2-n_1)}, \qquad (8.31)$$

where b has a positive real part. Inserting this in (8.27), some simple algebra gives the relations

$$\cos Ga = e^{-ba},$$
$$\epsilon = 2\gamma \sin^2 Ga. \qquad (8.32)$$

There is thus just one solution for each value of G; its energy is positive, but less by a factor $\cos^2 \tfrac{1}{2} Ga$ than the least energy of two separate spin waves whose wave vectors add up to $2G$; the latter by (8.26) is $8\gamma \sin^2 \tfrac{1}{2} Ga$.

The solution (8.31) has been called a 'spin complex' by Bethe (cf. Bethe and Sommerfeld, 1933). It represents a state in which the two wrong spins are bound together. Since adjacent spins have a preference for being parallel, it is plausible that it should take less energy to produce two wrong spins if they are kept together. But since there are altogether N possible values for G, and therefore N such bound states, they are statistically negligible compared to the 'free' spin-wave solutions (8.28), whose number is practically N^2.

For still smaller values of m, i.e. more wrong spins, we find again states with only free spin waves, which scatter each other on collision, and spin complexes of varying size which may also scatter each other and the single spins. Bethe has shown how to construct the general solution, but since it has not proved possible to generalize this solution to two or three dimensions we shall not discuss the details here.

All the solutions we have found have $\epsilon \geqslant 0$, i.e. energies not lower than that of the fully magnetized state $m = N$. This suggests that there exists no level with negative ϵ. This conjecture can be confirmed by the following argument. Suppose there were one level with negative ϵ. In the corresponding state function (8.11), choose the coefficient with the greatest modulus. This must exist, since the number of coefficients is finite. If several coefficients have equal modulus, we may choose any one of them. Let this belong to the arrangement $\{\mu^{(1)}\}$. Then (8.23) says that

$$A\{\mu^{(1)}\}\left[1 - \frac{\epsilon}{(N-r)\gamma}\right] = \bar{A}, \qquad (8.33)$$

where $N-r$ is, as before, the number of pairs of adjacent opposite spins, and $\bar{A}$ denotes the arithmetic mean of all coefficients occurring on the right-hand side of (8.23). If ϵ is negative, the factor on the left is greater than unity, and hence
$$|\bar{A}| > |A\{\mu^{(1)}\}|.$$

There must therefore be some coefficient which occurs in the average, and which exceeds $A\{\mu^{(1)}\}$ in modulus. This is a contradiction. We conclude that no state has lower energy than that with $m = N$. (The same argument shows that no state can have $\epsilon > 2N\gamma$.)

We can also find easily the mean energy of all states. A well-known theorem states that the mean of all eigenvalues of the set of equations (8.23) is equal to the diagonal sum of the coefficients on the right, which is γ times the mean value of $N-r$ over all spin arrangements. This is

$$\bar{\epsilon} = \tfrac{1}{2}N\gamma. \qquad (8.34)$$

8.3. Spin-wave model and ferromagnetism

We now turn to the case of a three-dimensional crystal, with one atom per cell, but otherwise keep all our simplifying assumptions.

It is then evident that we again obtain equation (8.23), where $\{\mu\}$ now means an arrangement of spins, each $\pm\tfrac{1}{2}$, over all lattice points, and $\{\mu'\}$ is an arrangement differing from it by the exchange of any two adjacent opposite spins. ϵ now has to be defined as

$$\epsilon = E - E_0 - zN(\alpha - \tfrac{1}{2}\gamma), \qquad (8.35)$$

where z is the number of neighbours which each atom has in the lattice. Again the state $m = N$ has $\epsilon = 0$, and the argument that no lower energy is possible applies directly. For one 'wrong' spin, we obtain in place of (8.26) a solution

$$A_{\mathbf{n}} = \text{constant} \times e^{i\mathbf{g}\cdot\mathbf{a}_{\mathbf{n}}}, \qquad (8.36)$$

with the energy $\quad \epsilon(\mathbf{g}) = \gamma \sum_{\mathbf{l}} (1 - \cos \mathbf{g}\cdot\mathbf{a}_{\mathbf{l}}), \qquad (8.37)$

§ 8.3　　　　　　　　FERROMAGNETISM　　　　　　　　171

the sum to be extended over the z lattice vectors $\mathbf{a}_l$ which join a given atom to its neighbours. For more than one 'wrong' spin we again have to consider combinations of spin waves which will scatter each other, and there may be spin complexes in which any number of wrong spins may be bound together.

Now we expect, and the calculations will confirm, that in statistical equilibrium at low temperatures most of the spins will be parallel, so that the number of wrong spins is a small fraction of N, and that the existing spin waves will all have small $\mathfrak{g}$. In that case we may neglect the scattering of the spin waves by each other, and also neglect the spin complexes.

That the latter is justified is not quite trivial, since the spin complexes represent, for each given total wave number, the state of lowest energy. However, we saw in (8.32) that for given total wave number $2G$ the energy of the spin complex differed from the state of two free spin waves by a factor $\cos^2 \tfrac{1}{2} Ga$, which for small G means a correction of higher order. This situation has been discussed for the one-dimensional case by Bethe.

We may then assume the spin waves independent, and introduce the number $n(\mathfrak{g})$ of spin waves of wave vector $\mathfrak{g}$ as variable. We may apply Bose-Einstein statistics to these and find in the usual way that

$$\bar{n}(\mathfrak{g}) = \frac{1}{e^{\epsilon(\mathfrak{g})/kT} - 1}, \qquad (8.38)$$

and for low temperatures we need to know $\epsilon(\mathfrak{g})$ for small $\mathfrak{g}$. In that case, expanding the cos in (8.37),

$$\epsilon = \tfrac{1}{2}\gamma \sum_l (\mathfrak{g} \cdot \mathbf{a}_l)^2,$$

which for cubic symmetry may be replaced by

$$\epsilon = \tfrac{1}{6}\gamma z d^2 g^2,$$

where d is the distance between nearest neighbours. For a simple, body-centred, or face-centred cubic lattice of cube side a this comes to

$$\gamma a^2 g^2. \qquad (8.39)$$

With this result, and using the fact that the density of permissible values of $\mathfrak{g}$ in reciprocal space is $1/(2\pi)^3$ per unit volume, we find for the total number of spin waves per unit volume

$$\frac{1}{(2\pi)^3} \int \frac{4\pi g^2 \, dg}{e^{\gamma a^2 g^2/kT} - 1}. \qquad (8.40)$$

When $kT \ll \gamma$, the integrand is small for g values comparable with the size of the basic cell, and we may therefore integrate to infinity. (At higher temperatures the approximations made previously would not be justified.) Then the proportion of 'wrong' spins is

$$\frac{1}{2\pi^2 \nu}\left(\frac{kT}{\gamma}\right)^{\frac{3}{2}} \int_0^\infty \frac{x^2\,dx}{e^{x^2}-1}, \qquad (8.41)$$

where ν is the number of atoms in a volume a^3, i.e. 1 for the simple cubic, 2 for the body-centred, and 4 for the face-centred lattice. The integral is 2·317 and we finally find for the magnetization at low temperatures

$$M = M_0\left[1 - \frac{0\cdot1174}{\nu}\left(\frac{kT}{\gamma}\right)^{\frac{3}{2}}\right]. \qquad (8.42)$$

This $T^{\frac{3}{2}}$ law seems in good agreement with the approach to saturation of real ferromagnetic metals at low temperatures. It is clear that our model does not apply precisely to those metals, since none of them has, at $T = 0$, exactly one Bohr magneton per atom. In Fe and Co, the number is nearer 2, and a more realistic approach here would be to assume two spins per atom. These will not be free to orient themselves independently, but will be linked by the interaction within one atom, which is stronger than the interatomic forces which we have considered. The atom therefore has a resultant spin of one unit, which has three possible orientations, its component in any direction being 1, 0, or -1. We must therefore consider 3^N states of the whole system. At temperature zero we may take all atoms in the state $m = 1$, and at low temperatures we must consider a few atoms in the state 0 and a few others in the state -1. A detailed discussion shows, however, that the completely reversed atoms ($m = -1$) are statistically negligible at low temperatures, and that for the others (8.36) and (8.37) still apply, so that the answer (8.42) for the low-temperature law is still correct (cf. Marshall, 1954).

A further complication is that we believe the magnetic electrons to be in $3d$ states of the atom, so that their wave functions are not isotropic. Such a state would normally have an orbital angular momentum as well as that due to spin. Since it is known from the gyromagnetic effect that the orbits play no part, we must assume that, because of the effect of the neighbouring atoms, the orbital moment of each atom is locked in a certain position absolutely, or relatively to the orbits of its neighbours, so that the orbits cannot respond to an external magnetic field.

Nevertheless, the wave functions will not be isotropic, and instead of a single exchange integral γ, we should really have to introduce several different quantities. No theory has yet been developed which would allow this fact to be taken into account.

It has also not proved possible to evaluate the magnetization at temperatures near the Curie point, and therefore we do not know whether the value of the Curie temperature would come out right if we took the value of the exchange integral from the low-temperature behaviour by (8.42). We also do not know whether the nature of the singularity in the magnetization curve or in the specific heat near the Curie point would be right.

It is possible to give the limiting behaviour at high temperatures far above the Curie point. Since in our model the number of states is finite, they become all equally probable at high temperatures, and the energy content becomes equal to their mean energy, which for one spin we have found to be $\frac{1}{2}N\gamma$. The value for two spins per atom can be found equally easily. Thus the integral over the magnetic specific heat (i.e. the observed specific heat corrected for the specific heat of the lattice) should at high temperatures reach a finite, and predictable, limit. In fact, the specific heat keeps rising even above the Curie point, and the amount is far too large to represent the specific heat of the $4s$ electrons. This shows up the limitation of the spin-wave picture; the electrons must have more freedom of movement than our simple model permits.

8.4. The collective electron model

The name 'collective electron' model is not very descriptive, since any theory of ferromagnetism must of necessity take the collective effect of many electrons as the basis, but it seems to have established itself as the name for the picture in which one starts from the problem of a single electron moving in a periodic field, as we have done in Chapter IV, and then considers the interaction between the electrons as a further modification.

In this picture we therefore start from a wave function like (8.10), in which, however, now the individual particle wave functions are not atomic functions, but solutions $\psi_k(\mathbf{r})$ for the periodic field. For simplicity we may consider only states from a given band, and thus omit the suffix l.

There are now, however, far more such functions to consider than before. For example, in the simplest case of N electrons, each in the periodic field of N atoms, we have to distribute N electrons amongst $2N$ spin and orbital states, and this gives $(2N)!/(N!)^2$ which is asymptotically 2^{2N}, instead of 2^N as in the spin-wave model. In addition, the states from which we start do not now all have the same energy. The mathematical problem arising from this model is therefore very much harder than the previous one, and very little is as yet known about the properties of its solutions.

Some indication can, however, be obtained from the following argument, which is due to Bloch (1929): We consider only wave functions which consist of a single determinant (8.10), and we restrict ourselves to the following two cases: (a) The electrons fill the $N/2$ lowest orbital states, with two electrons in each state, which must, of course, have opposite spins. This state has total spin zero, and we shall call this determinant function Φ_0. (b) The electrons fill the N lowest states, with one electron in each. Moreover, we assume all spins parallel. The total spin is $N/2$, and we shall call this determinant function Φ_M.

These functions are solutions of the Schrödinger equation of the system, except for the interaction between the electrons. To obtain the expectation value of the energy we must therefore add to the single-particle energy the diagonal element of the electron interaction.

In this case the single-particle functions are orthogonal, being different solutions of the same Schrödinger equation; hence the determinant function (8.10) is correctly normalized. This follows since after squaring the function, for each of the $N!$ terms in the first factor, there will only be one of the terms in the second factor that is not orthogonal to it, namely the one in which the same element is taken from each row. The square integral of each term in the determinant is evidently unity, so that the factor $1/N!$ is just right for normalization.

Now consider the diagonal element of

$$\frac{e^2}{r_{ij}}, \qquad (8.43)$$

where i and j refer to two electrons and r_{ij} is their distance. Consider a particular term from the first determinant. Now two terms from the second determinant may give non-vanishing integrals: the same term as in the first factor, in which the electrons i and j appear in the wave functions with wave vectors $\mathbf{k}$ and $\mathbf{k}'$, and have spins μ and μ' respectively, and that in which they have been interchanged. The first is

$$\alpha(\mathbf{k}, \mathbf{k}') = \int |\psi_\mathbf{k}(\mathbf{r})|^2 |\psi_{\mathbf{k}'}(\mathbf{r}')|^2 \frac{e^2}{|\mathbf{r}-\mathbf{r}'|} d^3r d^3r', \qquad (8.44)$$

and the second is the exchange integral

$$-\gamma(\mathbf{k}, \mathbf{k}') = \int \psi_\mathbf{k}^*(\mathbf{r})\psi_{\mathbf{k}'}^*(\mathbf{r}')\psi_\mathbf{k}(\mathbf{r}')\psi_{\mathbf{k}'}(\mathbf{r}) \frac{e^2}{|\mathbf{r}-\mathbf{r}'|} d^3r d^3r' \qquad (8.45)$$

if the spins μ, μ' are equal, and otherwise zero by the orthogonality of the wave functions.

§ 8.4 FERROMAGNETISM

Adding all terms together, and summing (8.43) over i and j, we find for the expectation value of the total energy in the case of the function Φ_0

$$2\{\sum E(\mathbf{k}) + \sum \alpha(\mathbf{k}, \mathbf{k}') - \tfrac{1}{2}\gamma(\mathbf{k}, \mathbf{k}')\}, \tag{8.46}$$

where the sum extends over values of $\mathbf{k}$ belonging to the $\tfrac{1}{2}N$ lowest eigenvalues $E(\mathbf{k})$. The second term represents the electrostatic interaction of the average electron density with itself, and should be omitted, since this has already been allowed for in the definition of the single-particle potential (cf. Chapter IV), and the energy of the state Φ_0 is therefore

$$2\{\sum E(\mathbf{k}) - \tfrac{1}{2} \sum \gamma(\mathbf{k}, \mathbf{k}')\}. \tag{8.47}$$

In the same way we find for the mean energy of Φ_M

$$\{\sum E(\mathbf{k}) - \tfrac{1}{2} \sum \gamma(\mathbf{k}, \mathbf{k}')\}, \tag{8.48}$$

the sums now extending over the N lowest states.

Evidently the first term in (8.48) is larger than that in (8.47), since we have moved electrons to states of higher energy. On the other hand, the other term will also be larger, since the sum now contains N^2 terms as compared to $N^2/4$ terms in (8.47), a change which should more than outweigh the factor 2 in front. Which of these two states gives lower mean energy depends therefore on the ratio of the exchange term to the single-particle energy.

For an idea of the magnitudes, one may take the case of free electrons. Then

$$\gamma(\mathbf{k}, \mathbf{k}') = \frac{4\pi^2 e^2}{V(\mathbf{k}-\mathbf{k}')^2}, \tag{8.49}$$

where V is the volume (cf. (6.65)). Hence (8.47) becomes

$$2\frac{V}{(2\pi)^3} \int_{|\mathbf{k}|<k_0} d^3\mathbf{k}\, \frac{\hbar^2}{2m} k^2 - \frac{2\pi^2 e^2 V}{(2\pi)^6} \int_{\substack{|\mathbf{k}|<k_0 \\ |\mathbf{k}'|<k_0}} \frac{d^3\mathbf{k}\, d^3\mathbf{k}'}{|\mathbf{k}-\mathbf{k}'|^2}, \tag{8.50}$$

where

$$k_0^3 = \left(\frac{3\pi^2 N}{V}\right). \tag{8.51}$$

After evaluating the integrals, the energy per unit volume is

$$\frac{\hbar^2}{10\pi m} k_0^5 - \frac{2e^2}{(2\pi)^3} k_0^4. \tag{8.52}$$

Similarly, for Φ_M,

$$\frac{\hbar^2}{20\pi m} 2^{\tfrac{2}{3}} k_0^5 - \frac{e^2}{(2\pi)^3} 2^{\tfrac{1}{3}} k_0^4. \tag{8.53}$$

Of these two expressions, the second is lower if

$$k_0 < \frac{5me^2}{2\pi^2\hbar^2} \frac{2^{\tfrac{1}{3}}-2}{2^{\tfrac{2}{3}}-2}. \tag{8.54}$$

It would therefore appear that the state with all electron spins parallel is the lowest state of the system if the electron density is low enough.

This result is no more than an indication of a trend, and an illustration of the fact that the exchange energy, which tends to set the spins parallel, increases more slowly with electron density than the kinetic energy of the Fermi gas, which favours opposite spins, so that ferromagnetism is favoured by a low electron density (small k_0). If one can generalize the result to electrons in a periodic field, (8.54) shows that a high effective mass favours ferromagnetism, which is in the right direction for the electrons in the narrow d-band of the actual ferromagnetic substances.

However, it is obvious that (8.52) and (8.53) must not be regarded as approximations to the correct position of the lowest energy level even for the hypothetical case of free electrons in the field of a uniform positive charge distribution. It is true that Φ_0 is the only state which has the lowest possible single-particle energy, and Φ_M is of all states of maximum spin the one with the lowest single-particle energy. But it is possible to construct very many other states of spin 0 or spin $\frac{1}{2}N$ by raising a few electrons to states just outside the edge of the Fermi sphere. This raises the single-particle energy by a very small amount, but in mixtures of such states we have freedom to make the exchange energy more negative. In fact, the matrix element (8.49) of the electron interaction is particularly large for states with very nearly the same momentum, and therefore the interaction energy is particularly sensitive to a mixture of states differing by very small changes in the electron momenta.

The importance of considering linear combinations of the determinantal function is also brought out by the fact that we must expect electrons with opposite spin, which are therefore not kept apart from each other by the Pauli principle, to be pushed from each other by the Coulomb force. This effect, which was mentioned in connexion with Wigner's work in § 5.2, represents a correlation between the electrons in coordinate space, and this can be achieved only by using linear combinations of products in momentum space. This effect reduces the electrostatic interaction, particularly in the case of low spin, and therefore tends to work against ferromagnetism.

The problem is even more difficult if we are not content to find out whether the state of lowest energy has a high spin value, but also want to find the magnetization as a function of the temperature. To do so, we must consider also the situation when not all the electrons occupying orbital states singly have their spins parallel. In that case we have, even for a given momentum distribution of the electrons, and for given total spin, many different determinantal wave functions, as in the spin-wave model.

Because of these mathematical difficulties no convincing solution of the problem has as yet been given. Nevertheless, the collective electron picture has certain advantages over the spin-wave model. It fits in naturally with the fact that the saturation moment is in general not a whole number of Bohr magnetons per atom, and it explains the rising specific heat well above the Curie point.

In the original paper by Bloch (1929) to which I have referred, the problem is simplified by assuming for each state a definite distribution of electrons in momentum space and then replacing the energy of all states of given spin, by their average. Stoner (1948) has given an extensive discussion of the consequences of this model.

8.5. Neutron scattering

It has been pointed out by Moorhouse (1951) that experiments on the scattering of very slow neutrons by ferromagnetic crystals would be capable of giving some rather detailed insight into the mechanism of ferromagnetism. Although suitable experiments are not easy and have not yet been carried out, it is of some interest to discuss this point as an amusing illustration.

The situation is very similar to that discussed in § 3.5. In a perfect crystal with complete translational symmetry, neutrons could not be scattered if their wave vector was less than the smallest non-zero vector of the reciprocal lattice. 'Wrong' spins, however, represent a deviation from this translational symmetry, just like the lattice waves considered in § 3.5. We have therefore to consider processes in which a 'wrong' spin is produced or absorbed. The first, which means that an electron which was in the direction of predominant magnetization is turned into the 'wrong' direction, with corresponding reversal of the neutron spin, requires energy and is therefore rare for slow neutrons. The second, in which the neutron turns a 'wrong' spin in the 'right' direction, may happen for very small neutron velocity, though it depends, of course, on the presence of 'wrong' spins and will therefore not happen at zero temperature.

We shall discuss this first using the spin-wave model.

For the absorption of a spin wave we have, by an obvious extension of (3.43) and (3.44),

$$\left. \begin{array}{c} \mathbf{k'} = \mathbf{k} + \mathfrak{g} + \mathbf{K} \\ \dfrac{\hbar^2}{2M} k'^2 = \dfrac{\hbar^2}{2M} k^2 + \epsilon(\mathfrak{g}) \end{array} \right\}. \qquad (8.55)$$

The spin-wave energy function $\epsilon(\mathfrak{g})$, which for the nearest-neighbour approximation is given by (8.37), is qualitatively of a similar form to the phonon energy function which appears in (3.44), except that it has a continuous derivative near the minimum as well as near the maximum. A one-dimensional representation of (8.55) would therefore be almost like Fig. 6, except that the 'comb' should be replaced by a cosine curve. This fact would widen the distance between the two points of intersection near each minimum, and therefore widen the peaks due to magnetic scattering as compared to the thermal peaks. On the other hand, the amplitude of the spin-wave curve is very much larger than that for phonons, which goes in the opposite direction. Moorhouse shows that, except for the peaks of lowest order, the spin waves give sharper peaks than the phonons.

For this estimate one has to know the value of the exchange integral γ; Moorhouse estimated this by extrapolating the low-temperature magnetization law (8.42) to the Curie point. This estimate has been improved by Marshall (1954), who pointed out that it is more consistent to fit (8.42) to the low-temperature magnetization measurements. In addition, he applied a spin-wave model with two spins per atom, which is nearer the correct value for iron. Both changes go in the direction of making the peaks still sharper.

The sharpness of successive peaks measures, in fact, the width of the spin-wave curve at different heights, and would therefore provide a detailed check on the spin-wave picture.

That these results are characteristic of the spin-wave model can be seen already from the fact that in the Weiss model the energy required to reverse a spin is assumed to depend only on the magnetization. In that case the scattered neutrons should all have the same energy, and presumably be distributed more or less uniformly in direction.

It is not easy to make a reliable prediction for the collective electron model. In the approximation in which the exchange energy is a function of the total magnetization, the spin reversal would imply moving an electron from the border region of the Fermi distribution of 'wrong electrons' to a place in the border region of the 'right electrons'. Near saturation the first Fermi distribution contains very few electrons, and we are near the origin in k space. The final point must have a momentum near k_0. The increased kinetic energy just outweighs the gain in exchange energy, since the two Fermi functions must be in equilibrium with each other. The energy change may therefore have all values of the order of kT, and it would have no strong correlation with the momentum transfer. This model therefore does not seem to lead to sharp peaks.

8.6. Remark on magnetization curves

The problem of spontaneous magnetization, with which this chapter has been concerned, represents only the raw material for most problems of practical interest. In subjecting a ferromagnetic material to an external field, one does not usually change the intensity of spontaneous magnetization in each small domain, but merely the orientation and the size of the different domains, unless one is fairly close to the Curie point or is dealing with extremely strong fields.

It is impossible, within the scope of this discussion, to do justice to this wide and important field, but a few remarks may help to show its connexion with the previous sections.

In the first place, it is of vital importance to take into account the fact that the spontaneous magnetization will be directed in one of a few crystallographically equivalent directions. Our previous formulae made the energy a function of the magnitude of the resultant spin, but since the spin came in only through the Pauli principle, only the relative spin direction of the electrons was important, not the absolute direction in space.

A directional dependence must therefore be due to magnetic interactions. The magnetic interaction between the electron spins cannot be responsible, at any rate in a cubic crystal, since it must be a quadratic function of the magnetization vector, and hence for cubic symmetry must be isotropic.

We do, however, find a suitable effect in the interaction of the spin (or spins) of each atom with the orbital motion. This effect is proportional to the product of the spin moment with the magnetic field due to the orbital motion. This magnetic field would vanish if it were not for the electron spin, since the evidence of the gyromagnetic effect shows that the orbital moments do not contribute appreciably to the magnetization, so that the different orientations of the orbit, which are degenerate in the atom, must be separated by the anisotropic field of adjacent atoms.

The spin will induce a moment in the atom, of which the leading term will therefore be proportional to the magnetic field of the spin, and, again by symmetry, have the same direction. To this order there is still no directional effect. However, in the strong local field due to the spin in the same atom, the susceptibility is not strictly constant. The field dependence of the atomic susceptibility is governed by the ratio between μH (where μ is the Bohr magneton) and the energy differences ΔE between the states corresponding to different orientations of the orbital wave function. These energy differences vanish for the free atom, and are expected to be reasonably small in the crystal. The H^2 terms in the atomic susceptibility (odd powers of H must go out by symmetry) are therefore not negligible.

In a single crystal it should therefore be easy to arrange all the domains in one of the directions of easy magnetization. To magnetize the crystal fully in a different direction requires, however, an external field strong enough to compete with the anisotropic part of the induced atomic field. Experimentally the magnetic fields necessary for this are of the order of a few hundred gauss, and this is quite compatible with the higher-order terms to which I have referred.

On the basis of the factors so far included, an infinitesimal field would be sufficient to magnetize a crystal up to its intrinsic magnetization in the direction of easy magnetization, and in any other direction up to the magnetization obtainable by turning all domains into that direction of easy magnetization which forms the smallest angle with the external field.

Hence we would expect an infinitely large initial susceptibility, and it is, in fact, correct that the initial susceptibility is extremely high for pure and well-grown crystals. To understand the actual susceptibility, and the hysteresis phenomena, one must consider the mechanism by which the domains are oriented. It is easy to see that this cannot take place by rotating the direction of magnetization gradually, since this involves passing through unfavourable directions which, in the absence of a strong external field, are not possible directions for the magnetization. What will happen instead is that domains with the correct orientation will grow at the expense of others, and this process depends on the nature of the boundaries between domains.

In an absolutely perfect crystal the boundaries would be able to move freely throughout the crystal, but, for example, a region of strain or impurities may reduce the exchange forces, so that at this point the spins have a smaller tendency to be parallel to each other, and a boundary passing through such a region contains less energy than if it passes through undisturbed material. The boundaries will therefore in practice be found in such positions of relative stability, and a magnetic field of finite strength is needed to move each boundary over the obstacles formed by less favourable positions. If the movement of boundaries is severely impeded, as in a quenched material with high internal stresses, one finds

§ 8.6 FERROMAGNETISM 181

high coercive forces, i.e. strong fields are required before an appreciable number of boundaries start moving.

Another fascinating problem is the speed of motion of the boundaries in an undisturbed region of the crystal.

8.7. Anti-ferromagnetism

Another important field which I shall not be able to cover adequately is the phenomenon of anti-ferromagnetism, which has been explored and described by Néel (1932, 1936).

Certain substances, including metals, alloys, and non-metallic crystals, show an anomaly in the specific heat at a temperature below which the magnetic susceptibility is smaller than above. Néel explained this as a tendency of adjacent atoms to have their spins opposite rather than parallel. In that case one would expect that, at sufficiently low temperature, the lattice could be subdivided into two sub-lattices, so that all or most of the spins of one sub-lattice were mutually parallel, but opposite to those of the other.

A large number of the properties can be described in terms of a Weiss model like that used in § 8.1, except that one has to assume that the force on each spin contains a term proportional to the total moment of the sub-lattice to which the spin belongs, and a term in the total moment of the other; since the tendency is towards opposing spins, the constant of proportionality in the second term must now be negative.

At first sight it might be expected that just such a behaviour would result if, in the spin-wave model of § 8.2, we assume the exchange integral γ to be negative. One might therefore expect that all substances for which the Heitler–London approximation of a fixed number of electrons in each atom is reasonably applicable should be either ferromagnetic or anti-ferromagnetic (unless γ happens to be zero). However, this is not borne out by a closer discussion of the model.

If in the formulae of § 8.2 we assume γ negative, then the state of lowest energy corresponds to the largest value of ϵ/γ. Now, whereas the smallest value of this ratio is easily found to be zero, and belongs to stationary states of a very simple kind, the highest value of ϵ/γ is not known. We have shown that it cannot exceed $2N$, but it is easy to see that it cannot, in fact, reach this value.

We have, in § 8.2, constructed a simple stationary state belonging to $\epsilon = 0$, and it corresponds very closely with the intuitive picture of a ferromagnetic state in which all spins are parallel. But the corresponding intuitive picture of adjacent spins being opposite, as in a chess board pattern, is not similarly a stationary state. One verifies fairly easily from (8.23), particularly in the form (8.33), that it is not a correct solution of this equation to assume $A = 1$ for the spin arrangement which has all adjacent spins opposite, and $A = 0$ for all others. The reason is quite clear: the exchange phenomenon does not allow spins to stay in their places, but they will change places with each other. This makes no difference (and is, in fact, unobservable) when they are all parallel but is important when they are not.

Because of these difficulties, no precise solution describing the highest values of ϵ/γ is available, and we do not know whether this model with negative γ would show the features of anti-ferromagnetism.

There is, however, little doubt that qualitatively the anti-ferromagnetism is due to an exchange interaction opposite in sign to the ferromagnetic case. It was at first regarded as a difficulty that the phenomenon was observed in oxides and other compounds in which the distance between the magnetic ions is so large that one could hardly expect the exchange interaction, which depends on the overlap of the atomic wave functions, to be at all appreciable. This was resolved by a suggestion due to Kramers (1934) of an indirect exchange involving several electrons, in which an intermediate non-magnetic atom takes part.

Terms representing such a multiple exchange are, in principle, contained in (8.12). They were neglected in deriving (8.21) since they contain higher powers of the 'overlap factor' (8.14), and in the ferromagnetic case they would have altered the results only in quantitative detail. However, as one knows from the binding energies of molecules, the overlap integral is not really small for realistic distances, and its square is certainly very much larger than the overlap between two magnetic ions which are two inter-atomic distances apart.

We may, for example, imagine that two magnetic ions each containing one electron are separated by a non-magnetic ion containing, in a closed shell, several electrons. Let u, w be wave functions in the two magnetic ions, and v_1, v_2 two different orbital states in the atom in the middle. We take all these with parallel spin, and number the electrons 1 to 4. Then the numerator of (8.12) contains a term of the kind

$$-u^*(x_1)v_1^*(x_2)v_2^*(x_4)w^*(x_3)\{V_{12}+V_{14}+V_{23}+V_{34}\}u(x_2)v_1(x_3)v_2(x_1)w(x_4)$$

(8.56)

integrated over all coordinates. The corresponding term is absent when the electrons in states u and w have opposite spin.

In current work on this subject it is usual to consider a slightly different mechanism of indirect exchange, which depends on the possibility of a virtual state in which the non-magnetic ion in the centre had lost or gained an electron from one of its neighbours. In the framework of the approximation outlined in § 8.2 such terms belong to an approximation of higher order (as does (8.56) because it contains four overlap factors) but in a realistic picture they need not, in fact, be small.

A summary of the current theory of anti-ferromagnetism can be found in a review by van Vleck (1951).

Notes added in Proof: It is worth pointing out that in the cases of one or two dimensions the spin-wave model gives, in place of (8.41), an integral in which $x^2 dx$ is replaced by dx or $x dx$ respectively. The integral then diverges at $x = 0$, showing that the approximations are not justified and that the proportion of 'wrong' spins is not small. Only the three-dimensional case, therefore, shows ferromagnetic behaviour.

The effect of the β^2 terms in (8.21) has been discussed in several papers. Carr (1953) comes to the conclusion that they may be neglected if one alters the meaning of the parameters α and γ to a slight extent.

Note added January 1956

W. Marshall (*Proc. Roy. Soc.* A, **232**, 48, 1955) gives a variation method to show that the simple model of this chapter, with interactions between adjacent magnetic atoms only, will not give an ordered state as a ground state for anti-ferromagnetism. It is conceivable that a more exact calculation would reverse this result, but in either case the energy difference between the lowest ordered and the lowest disordered state would be very small, so that the actual occurrence of anti-ferromagnetism is likely to depend on other small effects, such as the interactions between more distant atoms. This view seems quite compatible with the empirical evidence.

IX

INTERACTION OF LIGHT WITH ELECTRONS IN SOLIDS

9.1. General outline. Classical theory

IN this chapter we return to the action of light waves, which was also discussed in Chapter III. There we confined ourselves to the case in which, both before and after the transition, the electronic system was in its ground state. We must now discuss the more general case in which there are electronic transitions. This includes all optical problems in metals as well as the photoelectric effect. It also includes the light absorption and emission in all solids at frequencies comparable with atomic absorption lines. The exceptional cases of coherent scattering and vibrational Raman effect have already been treated in Chapter III.

Thus in metals one important class of processes, dominant at low frequencies, i.e. usually in the infra-red and visible regions, is concerned only with the properties of the ordinary conduction band. There are then still two important regions according to whether the light frequency is smaller or larger than the collision frequency $1/\tau$. When $\omega\tau \ll 1$, the effect of the field is essentially the same as that of a static field, so that the ordinary static conductivity is adequate for describing the behaviour of the metal. This is certainly the case for waves in the radio-frequency region, and in the far infra-red. Since the collision times for ordinary metals at room temperature are in the neighbourhood of $3 \cdot 10^{-14}$ sec., the light wavelength for which the period equals the collision time is of the order of 10^{-3} cm.

In the opposite limit the collisions are negligible and we are dealing essentially with electrons in a perfect lattice.

We can survey the important processes of this class by following the classical theory due to Drude (1904). We assume each electron to have a mean collision time τ defined as in § 6.1. Then the mean component of the electron velocity is given by

$$\frac{\partial \mathbf{v}}{\partial t} = \frac{e}{m}\mathbf{E} - \frac{1}{\tau}\mathbf{v}, \qquad (9.1)$$

where $\mathbf{E}$ is the component of the applied field. If the applied field depends on time through a factor $e^{i\omega t}$

(as usual the complex exponential serves mathematical convenience

and the real field is the real part of the final solution) it therefore follows that
$$\mathbf{v} = \frac{e\mathbf{E}}{m\{i\omega+(1/\tau)\}}. \tag{9.2}$$

Hence the current density due to all electrons is
$$\mathbf{j} = \frac{ne^2}{m\{i\omega+(1/\tau)\}}\mathbf{E} = \frac{\sigma_0}{1+i\omega\tau}\mathbf{E}, \tag{9.3}$$

where σ_0 is the conductivity for constant field (cf. (6.16)). Rationalizing the denominator,
$$\mathbf{j} = \frac{\sigma_0}{1+\omega^2\tau^2}(1-i\omega\tau)\mathbf{E} = \frac{\sigma_0}{1+\omega^2\tau^2}\left(\mathbf{E}-\tau\frac{\partial\mathbf{E}}{\partial t}\right). \tag{9.4}$$

A current which is proportional to the rate of change of the electric intensity, and hence out of phase with the electric field, is physically equivalent, not to a conduction current, but to the displacement current arising from a dielectric constant.

Hence the effective conductivity at frequency ω is
$$\sigma(\omega) = \frac{\sigma_0}{1+\omega^2\tau^2}, \tag{9.5}$$

and the effective dielectric constant is
$$K(\omega) = 1 - \frac{4\pi\tau\sigma_0}{1+\omega^2\tau^2}. \tag{9.6}$$

These relations show that, as long as $\omega\tau \ll 1$, the current is almost entirely in phase with the electric vector (as is evident from (9.4)), and therefore for most purposes the dielectric constant is unimportant. When $\omega\tau \gg 1$, most of the current is out of phase with $\mathbf{E}$, and the properties of an electromagnetic wave travelling through the medium are then mainly influenced by the dielectric constant, which, in this case, is approximately given by
$$K(\omega) = 1 - \frac{4\pi\sigma_0}{\omega^2\tau}.$$

This expression is independent of the collision time, since we introduced σ_0 merely as a convenient abbreviation for $ne^2\tau/m$, so that
$$K(\omega) = 1 - \frac{4\pi ne^2}{m\omega^2}. \tag{9.7}$$

For normal metals this quantity is negative even for frequencies considerably greater than the collision frequency. A negative dielectric constant means an imaginary refractive index, and therefore strongly damped waves.

The dielectric constant changes sign for a critical frequency

$$\omega_1 = \sqrt{\left(\frac{4\pi n e^2}{m}\right)}. \tag{9.8}$$

According to Zener (1933) this result may be linked with the observation that the alkali metals become transparent in the ultra-violet, and the magnitude of the frequency at which this begins is in reasonable agreement with (9.8) if one assumes one electron per atom, and equates m with the mass of a free electron.

In deriving these results we have neglected the effect of transitions between different bands, which is to be discussed later. In addition we have taken free electrons instead of electrons in a periodic field; this is likely to alter the results by numerical factors in cases in which the energy surfaces are not nearly spheres, or in which the effective mass is very different from the free electron mass. The general nature of the results is not likely to be sensitive to this approximation.

A further assumption is, however, the existence of a definite collision time. The result can be extended, without change, to the case of isotropic motion with a collision probability depending arbitrarily on the angle of deflexion. In that case the considerations used at the end of § 6.1 apply, and since the perturbing force has the same angular dependence as that due to a static external field, the appropriate collision time is again τ_1 (cf. (6.21)) and therefore the same as in the ordinary conductivity.

However, when we are dealing with an anisotropic energy surface, and the collisions are not such as to distribute the electrons uniformly over the energy surface after each collision, then the problem takes the form of an integral equation, in which the term arising from the time dependence varies over the energy surface in a different way from that due to the collisions, and therefore the answers depend on the ratio between the two terms, i.e. on $\omega\tau$. It is to be expected that the solutions can no longer be represented in the form (9.5), (9.6) with a constant value of τ.

The problem becomes even more involved at low temperatures, but even there it will still be true that with increasing frequency the current will change from a mainly conductive to a mainly dielectric phase, and the frequencies for which the transition occurs will again depend on the collisions. We saw in § 6.7 that the frequency of collisions was proportional to T^3, but that the mean angle of deflexion was proportional to T. Since collisions in which the electron is deflected by a small angle are unimportant also in the present problem, the effective collision frequency for this purpose is again proportional to T^5, as in the static conductivity. Hence one expects that (9.5), (9.6) should still be qualitatively correct, with the order of magnitude of τ given by the classical formula for the conductivity.

In experimental studies of the optical constants of metals one usually has to work with reflected light since the transmission is too small, except in the frequency region in which the dielectric constant (9.6) is positive but the transitions to higher bands have not yet started to play a part. In the case of reflection, one must be sure that the results are not influenced by the special properties of the surface. Apart from the purity and gross structure of the surface, which raise questions of experimental technique, one must remember that the electrons within a mean free path from the surface will behave differently from those inside the metal, since, in addition to ordinary collisions, they may be deflected by the surface itself. However, this effect is noticeable only over distances from the surface of the order of the distance an electron travels during the collision time (mean free path) or during a period of the light wave, whichever is the smaller. Since the electron velocity is very small compared to the light velocity, this distance is therefore much less than a wavelength, and it will in general be unimportant.

In using equation (9.2), it has been assumed that the field acting on each electron may be taken as the space average, which also appears in Maxwell's equations, and this is correct only if there is no correlation between the position of one electron and that of other electrons that contribute to the field acting on it. It is well known that in other circumstances, e.g. in calculating the dielectric constant of a non-conducting solid, in which each atom is in a highly symmetric position in relation to its neighbours, these correlations are important and give rise to the Lorentz-Lorenz factor.

If we are treating free electrons as strictly independent, there are, of course, no correlations. On the other hand, we know that both the Pauli principle and the electrostatic repulsion between the electrons tend to keep the electrons apart, so that each electron is, on the average, surrounded by a spherical region in which the chance of finding another electron is less than its mean over all space. It would appear plausible, therefore, that some part of the Lorentz-Lorenz correction should apply. On the other hand, Wilson (1936) concludes that no correction is necessary.

The existence of the periodic potential would seem to increase the correlations, and hence the case for a correction; certainly in the case of strong binding, each electron is likely to be found near some atom, and any other electrons that might be nearby are most likely to be found in the neighbouring atoms, and are very unlikely to occupy the same atom.

9.2. Transitions between bands

In transitions of an electron from one band to another, we have again to apply the by now familiar conservation of the wave vector. Strictly speaking, this should involve an equation between the initial and final wave vectors $\mathbf{k}$ and $\mathbf{k}'$ of the electron and that of the photon, but, unless we are dealing with X-rays, the photon wave vector is negligibly small, and it is a sufficient approximation to assume $\mathbf{k}' = \mathbf{k}$. Thus an electron in a given state $\mathbf{k}$ in the conduction band may make optical transitions only to a state with the same value of $\mathbf{k}$ in another band. However, the absorption frequency depends on $\mathbf{k}$; it is given by

$$\hbar\omega = E_{l'}(\mathbf{k}) - E_l(\mathbf{k}), \qquad (9.9)$$

where l and l' refer to the two bands. The variation of energy with $\mathbf{k}$ differs in different bands; for example, one band may have its lowest energy at $\mathbf{k} = 0$, and another its highest. There will, however, be a minimum value of (9.9), and therefore a minimum frequency below which this process cannot take place. However, this frequency is not easily interpreted since it depends on the nature of the energy functions for both bands.

Experimentally this absorption edge should be more pronounced if its frequency is higher than the frequency (9.8) at which the one-band part of the dielectric constant ceases to be negative, as in the case of the alkalis.

Formulae for the values of K and σ including the inter-band transitions can be found in the literature; since, however, little use is made of these results in practice, we shall not derive them here.

If we increase the light frequency, we shall be concerned with transitions of electrons from the conduction band to higher bands, and also with the transitions from the lower energy states into the conduction band or just above it. The first of these cases is not interesting, since it will show very little structure, because in this case the energy change of the electron is again given by (9.9) with both states in broad bands and the different upper bands overlapping more and more for higher energy.

On the other hand, the absorption of X-rays by the inner electrons gives very much clearer results, since we may regard the bands belonging to the X-ray levels of the atoms as having negligible width, so that in (9.9) $E_l(\mathbf{k})$ is practically constant. The frequency of the absorbed radiation is therefore directly the energy of the final state, apart from a constant.

The strength of the absorption is proportional to the density of states in the final band, multiplied by a transition probability. This is proportional to the square of the matrix element

$$\int \psi_{\mathbf{k}l}^{*}(\boldsymbol{\epsilon}\cdot\mathrm{grad})\psi_{\mathbf{k}l'}\,d^3r, \tag{9.10}$$

where $\boldsymbol{\epsilon}$ is the polarization vector of the radiation. The integral is to be taken over the whole crystal. However, by the Bloch theorem the integrand is periodic, so that, apart from factors, it is sufficient to integrate over the basic cell. In each basic cell $\psi_{\mathbf{k}l}$ is then essentially one of the inner atomic wave functions. Consider this matrix element first for $\mathbf{k} = 0$. Then we have seen from the Wigner-Seitz approximation (§ 5.2) that $\psi_{\mathbf{k}l'}$ is almost exactly isotropic. If therefore $\psi_{\mathbf{k}l}$ belongs to a p-level, the two factors in (9.10) have the same angular dependence, and therefore the integral does not vanish. In that case it does not depend on $\mathbf{k}$ very sensitively.

On the other hand, if the inner state is an s-state, then we have for $\mathbf{k} = 0$ an isotropic function multiplied by the gradient of another isotropic function, and the result vanishes upon integration over angles. In that case the matrix element vanishes for $\mathbf{k} = 0$; it must therefore be proportional to $\mathbf{k}$ near $\mathbf{k} = 0$, and the transition probability proportional to $\mathbf{k}^2$. Near the other end of the band, for large values of $\mathbf{k}$ the wave function may have to be a maximum on certain faces of the polyhedron that encloses the unit cell, and vanish on others; it has therefore more nearly the symmetry of a p-wave and the previous statements are reversed.

We expect therefore that the probability is reasonably constant throughout a band, except that it may decrease proportionally with the distance from one or the other end of the band, depending on where the minimum energy lies, and depending on the inner state involved.

The best results by this method are obtained using fairly soft X-rays, since with higher energies (very deep initial level) the width of the conduction band is a small fraction of the total energy difference. In this way Skinner (1938) and others obtained curves which give a clear idea of the level density as a function of energy in many metals.

The same method may be used with emission of soft X-rays, with the difference that the absorption method shows up only those levels which are normally empty and the emission depends on the upper level being occupied. The two results should supplement each other and it is found that the absorption curve has a sharp cut-off at its low-frequency end, and the emission curve extends from that frequency down.

The fact that these results agree, at any rate qualitatively, with the predictions of the simple theory is not at all a matter of course. As always, our model neglects the interaction between the electrons, except on the average. Now if an inner electron has been removed from one of the atoms, this must cause the potential distribution in that atom to be rather different from its neighbours. We shall discuss later a very similar effect in non-metals, where it will be found to be important. It would therefore be quite possible that in the neighbourhood of this particular atom the extra attractive potential would lead to an energy level for a conduction electron which is below the lower edge of the ordinary conduction band, and must then be a discrete level. If this were possible, one would expect to find, on the low-frequency side of the emission band, some discrete lines, or rather narrow peaks, since collisions may broaden these lines. This behaviour is not found in practice.

9.3. Photoelectric effect

A special case of light absorption is the photoelectric effect, in which the electron is able to leave the metal. If we apply to this (9.9), we find that the threshold for the effect is not given by the well-known equation

$$\hbar\omega = W, \qquad (9.11)$$

but has a higher frequency. This can be seen immediately in the following way. To realize the threshold value (9.11) the transition must lead from a state on the border of the Fermi distribution ($E = \eta$) to a free electron of velocity zero. Assuming, for simplicity, that the free surface is the plane $x = 0$, then the translational symmetry of the crystal in directions parallel to this surface ensures that the components k_y and k_z of the wave vector are unchanged as a particle passes through the surface. If its velocity is negligible outside, it must have had $k_y = k_z = 0$ inside. If the wave vector is conserved in the optical transitions, as assumed in (9.9), then the initial state must also have had $k_y = k_z = 0$. Now in each band the equation

$$E_l(k_x, 0, 0) = \eta \qquad (9.12)$$

has only two solutions, belonging to equal and opposite values of k_x, of which one will have the electron travelling away from the free surface. Hence the final energy $E_{l'}(\mathbf{k})$ is also a well-defined quantity for each. Of these values none will in general coincide with $W+\eta$, i.e. with the energy of an electron at rest outside the metal.

This argument ignores the importance of the metal surface for the motion of the electrons. The statement that in the optical transition the wave vector remains unchanged was derived using the translational symmetry of the crystal. Near a surface this translational symmetry in the direction normal to the surface is evidently lacking. Using the analogy between wave vector and momentum, we may say that the balance of momentum and energy which is impossible for a photon just above the threshold (9.11) can be restored by the surface forces, which may also take up momentum and transfer it to the crystal as a whole.

This effect is evidently of the greatest importance for electrons in a thin surface layer. In order to have the wave vector uncertain by an amount δk, the electron must be within a distance of about $2\pi/\delta k$ from the surface. One can therefore see that, unless the work function is very small, or one of the levels in the higher bands reached from (9.12) is nearly in the right place, the surface effect concerns a layer only a few atoms thick.

Since even in a strongly absorbing metal the light penetrates to a depth of 10^3 to 10^4 atomic distances, it is clear that, for the optical properties in general, the surface may be neglected. For the photoelectric effect it is of particular importance because electrons which have made the transition near the surface have a much greater chance of escaping than those starting from greater depth. If an electron has absorbed the phonon well inside the metal, it may make a collision before it reaches the surface. For this purpose the collisions with lattice waves or impurities are unimportant, since we know them to be almost elastic. They may deflect some electrons from the surface which would otherwise have escaped, but they also deflect some electrons towards the surface which previously travelled at an angle to it. However, the electron may make inelastic collisions with other electrons, and one such collision is in general enough to prevent its escape. Collisions between electrons were discussed in § 6.5 and found to be very infrequent, but this was for electrons which formed part of the Fermi distribution, and therefore had enough energy only for collision with a selected few other electrons in the border region.

In the present problem we are concerned with highly excited electrons, which may interchange energy with almost all the electrons in the Fermi distribution. In that case, the dimensional reasoning used in § 6.5 shows the mean free path to be comparable to the atomic distance, though probably larger numerically.

The following picture emerges: at frequencies just above the threshold

(9.11), only the 'surface photoelectric effect' is possible. At a higher frequency, which depends on the particular properties of the energy bands of the metal, a volume effect becomes possible in addition. This will in general be more intense than the surface effect, though not nearly by the same factor as one would estimate in the absence of inelastic collisions.

No detailed analysis of the volume effect seems to be available. The surface effect exists also in the absence of the atomic potential, and calculations for free electrons in the neighbourhood of a discontinuous or gradual potential rise near the surface give reasonable answers.

It has been pointed out by Mitchell (1934, 1936) that it is essential to take into account the change in the light wave near the surface due to the optical properties of the medium. This makes the problem somewhat too complicated for a detailed presentation here.

One interesting result is that the theory of free electrons near a plane surface predicts zero effect for a light wave which has its electric vector parallel to the surface (and hence always for normal incidence). The reason is that for free electrons the motion can be separated into the motion at right angles to the surface and that in a plane parallel to the surface. If the electric vector is parallel to the surface, the light wave does not influence the motion in the normal direction, and hence the wave vector in that direction cannot increase, as is required.

In fact, all metals do show an effect for any polarization or angle of incidence. This may be due to the fact that the electrons are not free, so that the motions in different directions are not separable, or to the effect of collisions with phonons, which are also capable of restoring the momentum balance. In addition, any deviation of the surface from the perfect plane which has been assumed, would tend to produce the same result.

9.4. Non-conducting crystals

The discussion of § 9.2 may also be applied to a case when the valency electrons just fill one band and the next one is empty. In that case the one-band transitions discussed in § 9.1 do not exist, and the substance is transparent up to an absorption edge in the visible or ultra-violet region. This absorption edge should correspond to a transition of an electron from the highest occupied band to the lowest empty band, usually in this case called the conduction band.

If the maximum energy in the lower band belongs to the same **k** as the minimum energy in the higher band, then the absorption edge should equal the width of the energy gap between the bands. It may well be that these two states belong to different values of **k**, and in that case the lowest absorption frequency compatible with (9.11) is larger than the width of the gap. However, as in the case of the photoelectric effect, there may be a small amount of absorption due to surface forces and other deviations which makes such transitions possible, and we must then expect the existence of a frequency range in which there is a weak absorption, which will also be sensitive to temperature and to the mechanical state of the crystal, followed by the strong absorption at a higher frequency.

The result of such an absorption process is one electron in the higher band, and one vacant place in the lower band. As long as we follow our usual approximation of neglecting the interaction between electrons, we should regard the 'hole' and the electron as moving independently. However, it is at once obvious that this is a much more doubtful assumption in this case than in the case of a metal. Indeed, the electron and the hole carry opposite electric charges and will therefore attract each other at all distances. In a metal, any excess charge will at once be neutralized by an adjustment in the density of conduction electrons, and hence two specific charges will influence each other only when very close together.

We must therefore expect not only states in which the electron in the upper band and the hole in the lower band separate and move independently, but also states in which they are bound together and revolve around each other in an orbit.

They may, in fact, remain in the same atom or group of atoms, in which case their description in terms of the band concept is not too convenient, and it is preferable to start from independent atoms rather than electrons moving singly in a periodic field. This means starting from the Heitler–London model of § 8.2, except that we now assume the ground state of each atom to be non-degenerate, and allow one or more atoms to be excited. The state in which one atom is excited then has an N-fold degeneracy, since the excitation of any one atom requires the same energy. This degeneracy is removed by the interaction between the atoms. The situation is precisely the same as for single electron states in the strong-binding approximation of § 4.2 or for the case of spin waves. The stationary states have again a well-defined wave vector and the energy is a function of this wave vector, so that we obtain a narrow band in place of each atomic excitation level. The width of this band is given by an interaction integral of the type

$$(\phi_0^*(1)\phi_1^*(2)W_{1,2}\phi_0(2)\phi_1(1)), \tag{9.13}$$

where ϕ_0, ϕ_1 are the wave functions for the ground state and an excited state of a complete atom, the arguments (1) and (2) refer to the co-ordinates of all electrons in atoms 1 and 2 respectively, $W_{1,2}$ is the interaction energy between all particles in atom 1 and those in atom 2, and the brackets again indicate integration over all coordinates. This expression is similar to the interaction integrals of § 4.2 or the exchange integral of § 8.2, except that it does not involve the bodily transfer of an electron from one atom to the other. It is therefore appreciable even where the atomic distance is so large that the wave functions do not overlap. If the atomic transition has a non-vanishing electric dipole moment associated with it, (9.13) decreases at large distances like the interaction between two dipoles, i.e. as the inverse cube of the distance between the atoms.

A state resulting from such an excitation travelling with a definite wave vector has been called an excitation wave or 'exciton' (Frenkel, 1936). It follows from our usual arguments that, since the crystal in the ground state has complete translational symmetry, the exciton formed by the absorption of a light wave in a perfect crystal must have a wave vector equal to that of the light wave, which for most purposes is negligible.

Hence, no matter how wide the exciton band, the absorption spectrum of a perfect crystal should contain only a sharp line for each atomic absorption line. These discrete lines belong to frequencies below the beginning of the continuous absorption discussed before.

We see therefore that, as the result of the absorption of a photon, we may be concerned with three different states. (*a*) An electron is excited, but remains within the same atom; the excitation may then be passed along to other atoms in the form of an excitation wave, but each atom remains neutral throughout. (*b*) The electron leaves its atom, but remains in its neighbourhood, forming a kind of 'hydrogen atom' together with the positive hole which it left behind. This bound pair may again travel through the lattice together, though in a perfect lattice light absorption leads only to states in which the wave vector associated with the motion of the centre of gravity of the pair is zero. (*c*) The electron and the hole may separate and move independently with equal and opposite wave vectors.

The first state may not always be possible. For example, in an ionic crystal like NaCl, the least strongly bound electrons are those in the negative chlorine ion. This has no atomic excited states, and the result of a transition of such an electron is always its removal from the parent ion. Sodium ions do have excited states, but their excitation energy is higher than that for removing an electron from the Cl⁻, and therefore the discrete excitation levels overlap with a continuous spectrum and are broadened to a sufficient extent to lose their clear identity.

Another cause of broadening is the motion of the atoms in the lattice. This point is best explained by reference to Fig. 16, which shows schematically the electronic energy of the ground state and the excited electron state as a function of some coordinate determining the positions of the atoms. For example, if we think of the transfer of an electron from a Cl⁻ ion to an adjacent Na⁺ ion in NaCl, the abscissa of the figure might be the distance between these particular Cl and Na nuclei.

The point A, which represents the equilibrium position for the normal lattice, is no longer that for the excited state, since the attractive force between the two ions has been reduced. Because of thermal vibrations and zero-point motion, the initial state does not correspond exactly to A, but to a small spread of positions in its neighbourhood, indicated by $a_1 a_2$. It is evident that a transition which takes place when the atoms are in their extreme positions causes an energy change indicated by the vertical broken lines, and that the variation of this is much larger than the thermal energy itself, which is measured by the height of the lower curve at a_1 or a_2 above the minimum.

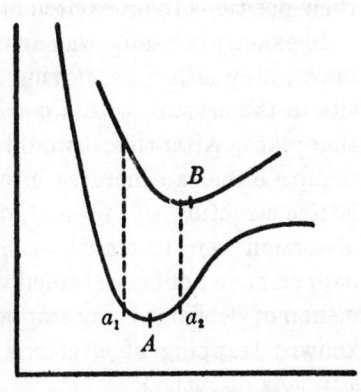

Fig. 16.

The use of this diagram is similar to the theory of molecular spectra. However, whereas in the problem of diatomic molecules there is only one coordinate, namely the distance of the two nuclei, the crystal has a very large number of degrees of freedom. Hence, for example, in the case of a molecule the curves of Fig. 16 would indicate that, after an absorption at or near A, the nuclei would not be in equilibrium and would be accelerated away from each other, the result being vibration with large amplitude, or dissociation, according to the shape of the curve.

In the crystal the vibration of one particular group of atoms must be strongly damped, since the energy will spread over neighbouring atoms, and the atoms should settle down near the new minimum B if such a point exists. The result is then a deformation of the lattice near the excited site, and then the excitation must be localized, and thus be a superposition of exciton states of all conceivable wave numbers. It is therefore plausible that this description applies only if the interaction (9.13) is weak enough. If it is strong, we must expect that the states have the structure of excitation waves whose location is undetermined, so that there is no inducement for any particular set of atoms to adapt their positions to an excited state.

In exactly the same way an electron in the upper band, or a hole in the lower, may either be moving freely, or may attach itself to a definite site in the crystal, with a corresponding adjustment of the lattice near that place. After this, it would no longer be free to move since this would require either an increase in energy, if it moved to a place where the lattice structure was normal, or a movement of the heavy nuclei, if the distortion were to travel with the electron. The conditions for this to happen have not been studied very extensively. Empirically this phenomenon of electrons being trapped in a uniform lattice does not seem to be known; trapping of electrons always takes place at sites where the lattice is disturbed.

The process of emission of light is, of course, in principle just the inverse of absorption; however, the main question here is how the excited state is produced and what other processes compete with the light emission. These problems belong to the material of the next chapter.

X
SEMI-CONDUCTORS AND LUMINESCENCE

10.1. Semi-conductors

IN Chapter IV we divided solids into two classes, those in which, in the state of lowest energy, the highest occupied state was the top end of a band, and those in which the last occupied level was somewhere inside a band. This is the distinction between insulators and metals. Since the definition refers to the state of lowest energy, it is appropriate to the absolute zero of temperature, and is then a sharp division. However, if the energy gap between the last full and the next empty band is small, a few electrons will be moved up by thermal agitation, so that there are a few conduction electrons, and a few holes in the lower band.

In such substances the number of carriers should increase very rapidly with temperature. They are characterized by a low electric conductivity which increases with temperature, and are called semi-conductors. The type I have described, in which temperature causes a transition between the bands of a perfect crystal, is known as an intrinsic semi-conductor.

This type is fairly rare, since it requires a rather small value of the energy gap to give a reasonable conductivity at, say, room temperature. In principle, any non-conducting crystal should be a semi-conductor at sufficiently high temperature, but usually the effect is quite negligible.

The substance may also contain impurities in which there exists an occupied electron state, or a vacant level, in the gap between the highest filled and the lowest empty state of the crystal. If the occupied level lies near the upper edge of the gap, a small addition of energy is sufficient to raise an electron into the conduction band; if an empty level lies near the lower edge of the gap, a small addition of energy will raise an electron from the occupied band to the impurity level, and leave a hole in the band. These two cases are referred to as 'donor' and 'acceptor' levels respectively.

Such impurities need not consist of foreign atoms; they may simply consist of an excess or deficiency of the atoms making up the normal lattice. For example, a missing Cl ion in a rock salt crystal causes an excess positive charge in the neighbourhood, and this neighbourhood must therefore contain electron levels which are lower than the unoccupied levels of the main crystal.

Similarly one finds in oxides, particularly those which exist chemically in different stages of oxidation, that the oxygen content of the crystal may differ from that of the perfect crystal, and any vacant lattice sites or additional ('interstitial') atoms count as impurities for the present purpose.

At first sight it seems surprising that also in impurity semi-conductors the conductivity should always disappear at low temperatures, since there ought to be cases in which a donor level in the impurity atom lies higher than the bottom edge of the empty band, or an acceptor level below the upper edge of the filled band. In such a case one would expect the number of carriers to be equal to that of suitable impurity atoms and independent of temperature.

The answer to this apparent difficulty lies again in the fact that we must not for this purpose neglect the electron interaction. If, in a neutral lattice, an extra atom is inserted which loses an electron to an empty band of the lattice, the atom will be left ionized, and therefore attract the electron. The state of lowest energy, and therefore the only one realized at zero temperature, is therefore an orbit of the electron bound to the positively charged centre. An electron in such a bound orbit is, of course, not free to take part in the conduction process, and hence a finite energy is required to release it for this purpose.

The same states in all probability exist also in intrinsic semi-conductors, so that the production of electrons and holes which are bound together requires somewhat less energy than the state in which they are separated. Only the particles in the latter state act as carriers of electricity, and its energy determines the temperature dependence of the number of carriers in the semi-conductor.

10.2. Number of carriers

To calculate the number of carriers in an intrinsic semi-conductor, we assume that E_1 is the top of the occupied band, E_2 the bottom of the empty band. $Z_1(E_1-E)$ is the number of states in the lower band above energy E, and $Z_2(E-E_2)$ the number of states in the upper band below E.

Then, using Fermi's distribution, and allowing for spin, we have

$$n_1 = 2 \int \left|\frac{dZ_1}{dE}\right| \frac{dE}{1+e^{-\beta(E-\eta)}}; \quad n_2 = 2 \int \frac{dZ_2}{dE} \frac{dE}{e^{\beta(E-\eta)}+1}, \quad (10.1)$$

where n_1 and n_2 are the numbers of holes in the lower, and of electrons in the upper band, respectively. These two quantities must be equal if the crystal is to be neutral, and at moderate temperatures both will be small. To make both expressions small, both denominators must be large, and therefore the 1 can be neglected. We then have

$$\left.\begin{array}{l}n_1 = 2e^{-\beta(\eta-E_1)} \displaystyle\int_0^\infty \frac{dZ_1(\epsilon)}{d\epsilon} e^{-\beta\epsilon}\, d\epsilon = e^{-\beta(\eta-E_1)} K_1 \\[2mm] n_2 = 2e^{-\beta(E_2-\eta)} \displaystyle\int_0^\infty \frac{dZ_2(\epsilon)}{d\epsilon} e^{-\beta\epsilon}\, d\epsilon = e^{-\beta(E_2-\eta)} K_2\end{array}\right\}, \quad (10.2)$$

where we have used in the first integral $E_1 - E$, and in the second $E - E_2$ as variable of integration.

If we may assume the energy to be a quadratic function of the wave number near the maximum or minimum energy in either band, we may characterize the states in each by an effective mass (this has to be a suitable average over directions if the energy is anisotropic near the maximum or minimum), so that, per unit volume,

$$K_1 = \frac{1}{4}\left(\frac{2m_1 kT}{\pi \hbar^2}\right)^{\frac{3}{2}}, \qquad K_2 = \frac{1}{4}\left(\frac{2m_2 kT}{\pi \hbar^2}\right)^{\frac{3}{2}}, \qquad (10.3)$$

m_1 and m_2 being the effective masses. This shows that, to make n_1 and n_2 equal,

$$\eta = \tfrac{1}{2}(E_2 + E_1) + \tfrac{3}{4} kT \log \frac{m_1}{m_2}. \qquad (10.4)$$

Our approximation assumes that the gap $E_2 - E_1$ is much larger than kT, and, unless the masses are very different indeed, the second term is negligible. The Fermi energy η then lies just in the middle of the gap, and it is now easy to verify that the neglect of the 1 in both denominators of (10.1) is justified. Inserting in (10.2),

$$n_1 = n_2 = \frac{1}{4}\left(\frac{2m^* kT}{\pi \hbar^2}\right)^{\frac{3}{2}} e^{-(E_2 - E_1)/2kT}, \qquad (10.5)$$

where
$$m^* = \sqrt{(m_1 m_2)}. \qquad (10.6)$$

Equation (10.5) shows the characteristic exponential variation of the number of carriers with temperature. The occurrence of one-half the required energy in the exponent is typical of dissociation reactions.

The electric conductivity depends on the number of carriers and their collision time, but the latter will be a slowly varying function of temperature compared to the exponential in (10.5), and therefore in general the law (10.5) also gives the variation of the conductivity with temperature.

In an impurity semi-conductor, which has n_d donor atoms per unit volume, each containing one electron in an energy level E_d, the second equation (10.2) is still valid, but the first has to be replaced by

$$n_1 = n_d e^{-\beta(\eta - E_d)}, \qquad (10.7)$$

where n_1 is now the number of vacant donor levels, which must again equal n_2.

Then
$$\eta = \tfrac{1}{2}(E_d + E_2) + \tfrac{1}{2}kT \log(n_d/K_2). \qquad (10.8)$$

Hence
$$n_2 = \tfrac{1}{2}\sqrt{n_d}\left(\frac{2m_2 kT}{\pi \hbar^2}\right)^{\tfrac{3}{4}} e^{-(E_2 - E_d)/2kT}. \qquad (10.9)$$

The important features of this formula are again the exponential variation with temperature, and the proportionality to the square root of the concentration of donor atoms.

An analogous formula holds, of course, for an acceptor level.

If there are at the same time donor and acceptor atoms and the donor levels are higher than the acceptor levels, electrons leave the donor atoms and go to the acceptors, until either type is exhausted. Then (10.9) must be modified. If the acceptor level lies above the donor level, the Fermi energy lies between them, and hence either the number of electrons in the upper band or the number of holes in the lower is less than it would be in the absence of either kind of foreign atom.

However, in the same substance, different kinds of foreign atoms may in varying conditions produce either predominantly excess electrons, or predominantly vacant places.

10.3. Electrical properties

Once the number of carriers is given, we can discuss the properties of the substance in the same way as was done in Chapter VI for a metal. One important difference is that the electron density is generally small, so that the electrons or holes do not form a degenerate Fermi gas, and the equations of Boltzmann statistics apply instead. The small number means also that some of the properties of the electrons, such as their contribution to thermal conductivity, or to the specific heat, are swamped by the contributions from the lattice vibrations, and are of no great practical interest.

As regards the conductivity, the considerations of Chapter VI apply in general. If we are dealing with a cubic crystal, and if the energy surface of the empty band has only one minimum (or that of the full band only one maximum) the energy function $E(\mathbf{k})$ is isotropic. Collisions with impurities will, on the average, also be isotropic, since the electron waves are long, and the crystallographic anisotropy of the scattering centres is therefore unimportant. One may therefore apply the collision time concept with reasonable confidence.

Collisions with phonons are also approximately isotropic, though the anisotropy of the elastic constants makes the sound velocity depend on the direction of travel of the phonon relative to the crystal axes. It is also still possible to regard the collisions as elastic, even though the kinetic energy of the electrons is now only of the order of kT. This is because the most energetic phonon which a given electron can absorb is one which will just reverse the direction of motion of the electron. Conservation of wave vector and of energy then require that

$$f = 2k + \frac{2m^*c}{\hbar}, \qquad (10.10)$$

where m^* is the effective mass and c the sound velocity. The phonon energy is then small compared to the initial energy of the electron, provided
$$4m^*c \ll \hbar k, \qquad (10.11)$$
in other words, provided the electron velocity is large compared to the velocity of sound. The mean thermal velocity at room temperature, assuming $m^* \sim m$, is about 10^7 cm./sec. This is large compared to c, and therefore the energy transfer in a collision is negligible.

Hence we may apply the collision-time concept. The collision time of an electron may be obtained from (6.49), neglecting $n(\mathbf{k}, l)$ and $n(\mathbf{k}',l')$ compared to unity, and assuming $n(\mathbf{k}, l)$ to contain a term proportional to the cosine of the angle between $\mathbf{k}$ and a fixed direction. Subject to the condition (10.10), we may then neglect the phonon frequency. For fixed $\mathbf{k}$ the temperature enters only through the phonon numbers, and if the temperature is not too low, so that the quantum properties of phonons of wave number less than $2\mathbf{k}$ are still not important, N and $N+1$ are proportional to the temperature. The collision frequency for a given electron state is thus still proportional to T.

As regards the dependence on $\mathbf{k}$, we note that the summation over $\mathbf{k}'$ may be replaced by an integration, the volume element in k space being $k'^2\, dk'd\Omega'$, where $d\Omega'$ is the element of solid angle. Since the differential of the energy is $dE' = \dfrac{\hbar^2}{m^*} k'\, dk'$, we are left with a factor k after removing the δ-function by integrating over energy. The square of the matrix element $|(\mathbf{k}',l'|A|\mathbf{k},l)|^2$ is by (6.61) proportional to the phonon frequency. On the other hand, the phonon number N is, by (2.4), inversely proportional to the phonon frequency in the classical limit. It therefore follows that the collision frequency is proportional to the wave vector, or to the velocity of the electron, and this may be expressed by saying that the mean free path is independent of the velocity.

As regards the Hall effect, the existence of a definite collision time makes the theory of § 7.4 applicable, as far as (7.58). Since, however, the electrons do not all have substantially the same velocity, (7.59) must be replaced by a more careful calculation of J_y from (7.58) and (7.57). One then sees that in (7.59) τ is replaced by

$$\overline{\tau^2 v^2}/\overline{\tau v^2},$$

where v is again the electron velocity, and the average is to be taken with a Maxwell distribution, while in the conductivity (6.16) the appropriate mean value is

$$\overline{\tau v^2}/\overline{v^2}.$$

Hence, using the result that $v\tau$ is independent of velocity, (7.61) must be multiplied by the ratio

$$\overline{v^2}/(\bar{v})^2,$$

which for a Maxwell distribution is $3\pi/8$. Hence the Hall coefficient of a semi-conductor in which only electrons or holes contribute, and in which there is isotropy, is

$$R = \frac{3\pi}{8ecN}. \qquad (10.12)$$

One must, however, be certain that the conditions for the validity of this equation are satisfied before using it to estimate the number of electrons or holes. In particular in an intrinsic semi-conductor, in which the electrons and holes generally give comparable contributions, one must replace the equation by (7.65), and this usually cannot be evaluated from the observations without further assumptions about the model.

10.4. Density gradients and space charge (cf. Mott, 1938)

So far we have been concerned only with cases in which the electron density is uniform in space. In semi-conductors one is also interested in the case in which the density is non-uniform, either because the density of impurity atoms which act as donors or acceptors is non-uniform, or because electrons are raised to the upper band by a light wave which either has an intensity gradient because of the absorption of the substance or, because of its direction, gives a preferential direction to the electron velocity. I shall discuss only the first of these alternatives in detail; the other is of importance in connexion with the contact layer photoelectric effect, which will not be treated.

Suppose that a region in the crystal contains a density n_d of donor levels per unit volume, so that the equilibrium density of electrons would be given by (10.9), and suppose that the energy gap between filled and empty band is much larger than kT, so that the 'intrinsic' effect is negligible. Suppose also that in the neighbourhood there are other sources of electrons (such as a region with more donor atoms, or a contact with a substance of high electron density and low work function or the reverse) so that the electron density will differ from the equilibrium value. Within the given region, we may still use the second equation (10.2) for the electron density, and (10.7) for the number of vacant donor levels, but we may no longer require the two to be equal. There will then in general be a space-charge density, given by

$$e(n_2-n_1), \tag{10.13}$$

and this will cause an electrostatic potential $\phi(\mathbf{r})$ which has to be allowed for in discussing electronic levels. This potential will vary by a negligible amount over atomic dimensions, and we may therefore simply replace the energies E_2 and E_d by $E_2-e\phi$ and $E_d-e\phi$, respectively. From Poisson's equation and (10.13), assuming ϕ to depend only on one coordinate x,

$$\frac{d^2\phi}{dx^2} = -4\pi e(n_2-n_1). \tag{10.14}$$

For equilibrium the parameter η must still be a constant, independent of x. Inserting for n_2 and n_1 from the second equation (10.2) and (10.7),

$$\frac{d^2\phi}{dx^2} = -4\pi e n_2^0 (e^{e\phi/kT} - e^{-e\phi/kT}), \tag{10.15}$$

where n_2^0 is the equilibrium electron density, which is given by (10.9). It might be objected that in obtaining (10.15) we have used the value (10.8) for η, which would seem arbitrary; however, a change in η is equivalent to adding a constant to the electric potential ϕ, which is compatible with (10.14). Taking η equal to the equilibrium value amounts to the convention that $\phi = 0$ where the electron density has its equilibrium value.

The solution of (10.15) for which $d\phi/dx = 0$ when $\phi = 0$ is

$$e\phi = 2kT \log \coth\left(\frac{x-x_0}{l}\right), \tag{10.16}$$

where

$$l^2 = \frac{kT}{2\pi e^2 n_2^0} \tag{10.17}$$

and x_0 is arbitrary. Hence for small deviations from equilibrium, when ϕ is small,

$$e\phi = 4kT e^{-2(x-x_0)/l}, \tag{10.18}$$

so that to the same accuracy

$$n_2 - n_2^0 = \text{constant} \times e^{-2x/l}. \qquad (10.19)$$

On the other hand, if the source gives rise locally to a density greatly exceeding n_2^0, so that $e\phi$ is much larger than kT, we have

$$n_2 = n_2^0 e^{e\phi/kT} = n_2^0 \frac{l^2}{(x-x_0)^2} = \frac{kT}{2\pi e^2 (x-x_0)^2}. \qquad (10.20)$$

The last form shows that the result is now independent of n_2^0 and the formula applies therefore also when there are no donor levels present. The result may be expressed by saying that $1/n_2$ increases linearly with x.

We conclude therefore that small deviations from the normal electron density are evened out over a distance of the order (10.17), which for low values of the equilibrium density may be very large compared to the lattice spacing. It is also obvious why it is not worth considering the corresponding problem for a metal, since then the electron density is high and deviations are adjusted almost at once. Our formulae are, of course, not valid for metals, since the electron degeneracy has to be taken into account, so that the results contain η rather than kT, but the length that then replaces l is again very small indeed.

If, instead of static conditions, we assume the presence of an external potential gradient, there will be an electric current J of the form

$$J = -\sigma \frac{d\phi}{dx} - D \frac{dn}{dx}, \qquad (10.21)$$

where σ is the conductivity, and D is a diffusion coefficient. We know from the preceding discussion that the electrons must be in equilibrium, and hence the current zero, if

$$n_2 = \text{constant} \times e^{-e\phi/kT},$$

and this shows that
$$D = \frac{kT}{e} \frac{\sigma}{n_2} = kTu, \qquad (10.22)$$

a relation due to Einstein. $u = \sigma/n_2 e$ is independent of the electron density, and is called the mobility.

In the most general case of non-uniform electron density and an electric current, (10.21), in which for steady conditions J must be constant, i.e. independent of x, should be combined with the Poisson equation (10.14). This situation is, however, too general for a simple analysis, and we shall consider in the next section only a particular application.

10.5. Rectifying contacts

An important phenomenon involving semi-conductors is the rectifying contact. This is a contact between a metal and a semi-conductor which has, for zero current, a high resistance to flow across the surface, and in which the resistance diminishes rapidly with increasing voltage in one direction, the direction of easy flow, whereas for a potential difference of opposite sign the resistance increases strongly, at least over a certain range.

Several possible mechanisms are known which could give rise to a dependence of the contact resistance on the potential difference. We shall here discuss only one particular model, the theory of which is due to Mott (1939), and which is well borne out by the experiments in the case of copper-cuprous oxide contacts.

The high resistance of the contact is then due to a poorly conducting layer ('blocking layer') and it is known that in many cases this is a layer in which the impurity atoms are lacking. An important type of contact is that in which a metal surface is oxidized. On one side we then have an oxide in which excess oxygen is the activating impurity; it is reasonable that where the oxide is in contact with pure metal the excess oxygen should be lacking.

Suppose now for uniqueness that the carriers are electrons rather than holes and suppose that the work function of the metal exceeds that of the normal semi-conductor. Then there must exist a contact potential between the two layers, which is produced by the metal absorbing a few electrons from the semi-conductor. However, these electrons cannot come from the blocking layer, which contains practically no electrons, and must therefore come from the 'good' semi-conductor beyond it. In other words, the double layer which produces the contact potential has a thickness equal to that of the blocking layer. Instead of being restricted to a few atomic layers, the rise of the potential is thus spread over a layer which, in typical cases, has been found to be 10^{-4} to 10^{-3} cm. thick.

This state of affairs is shown in Fig. 17 (a). The Fermi level of the metal is at the same energy as that of the semi-conductor, which lies half-way between the bottom of the conducting band and the donor levels. The electrons from the donor levels penetrate somewhat into the blocking layer, but the rising potential makes their density fall rapidly.

If now a field is applied in such a direction as to pull electrons towards the semi-conductor, the situation will be as indicated in Fig. 17 (b). The main part of the additional potential difference resides in the blocking layer, and helps to make the gradient steeper. This diminishes the number of electrons which come, by thermal energy fluctuations, from the semi-conductor, and does not appreciably assist the electrons from the metal in reaching the blocking layer. It is only when the field becomes strong enough to give the conditions for 'cold emission' (cf. § 4.7) that the resistance will again diminish.

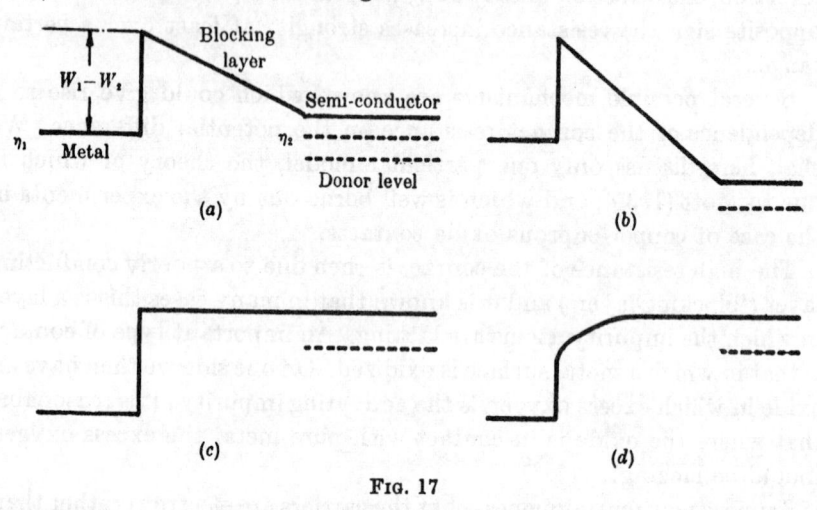

FIG. 17

On the other hand, if the sign of the applied potential difference is such as to pull electrons towards the metal, the potential distribution will approach that of Fig. 17 (c), with the potential in the blocking layer practically constant, so that the electron density will be the same as in the presence of the donor levels; the resistance of the blocking layer is then negligible.

To be precise, a high conductivity of the blocking layer means a high electron density, and therefore a space charge. We ought therefore to think of the limiting case in terms of Fig. 17 (d), in which the curvature of the potential in the blocking layer is due to the space charge.

The treatment by Mott neglects this space-charge effect, and assumes therefore that the potential gradient in the blocking layer is uniform. Then, from (10.21),

$$J = u\left(neF - kT\frac{dn}{dx}\right), \tag{10.23}$$

where now u, F, and J may be regarded as constant. I have omitted the suffix 2 on n. The integral of (10.23) is

$$n = \frac{J}{euF} + Ce^{-eFx/kT}. \tag{10.24}$$

The electron density is given at the metal surface, where it equals the first part of (10.2), with the value of η adjusted to that in the metal. It is also known for $x = d$, in the normal semi-conductor.

Then
$$J = euF \frac{e^{eFd/kT}n(0) - n(d)}{e^{eFd/kT} - 1}. \quad (10.25)$$

If the exponent is not a large positive number, the electron density will be large throughout the blocking layer, and the resistance negligible. If the exponent is large we neglect the 1 in the denominator,

$$J = euF\{n(0) - n(d)e^{-eFd/kT}\}. \quad (10.26)$$

Remembering that, in the absence of an external potential difference, as in Fig. 17 (a),
$$\frac{n(d)}{n(0)} = e^{eF_0 d/kT},$$

we have
$$J = \frac{eun(0)}{d}(V_0 - V)(1 - e^{eV/kT}), \quad (10.27)$$

where V is the applied potential difference and

$$V_0 = F_0 d.$$

(10.27), which is valid only for $V < V_0$, shows the characteristic unsymmetric behaviour of a contact rectifier.

This treatment depends on the thickness of the blocking layer being large compared to the mean free path of the electrons, since otherwise higher derivatives of the density would appear in (10.23). It also neglects space charges, as I have pointed out. It can, of course, be immediately extended to hole conduction, with the appropriate changes in sign.

10.6. Electrons not in thermal equilibrium

In the preceding sections, we were concerned with thermal equilibrium, so that the number of electrons in the upper band, or holes in the lower band, was in each case a constant, independent of time, except for fluctuations. We are then in general not interested in the mechanism by which the electrons are raised, since the result follows from statistical mechanics, independently of the mechanism.

The situation is different when the electrons are raised into the upper band by an external agent, such as a light beam, and we are then concerned with the mechanism, and also with the subsequent fate of the electrons. To fix our ideas, let us consider a case in which an electron is raised to the upper band by absorption of light from a state in the filled band.

In this case the result is an electron in the upper and hole in the lower band. There will be an electrostatic attraction between the two, and, if their kinetic energy is low, they may remain in a bound state, in which they revolve about each other, losing energy by various processes fairly rapidly and ending up after a short time with emission of light of a frequency slightly lower than that of the incident radiation and the return of the electron to the lower band. If the energy of the transition is high enough, the electron and hole can overcome their attraction and separate. Each then travels independently of the other, but is subject to changes of direction, due to collisions with lattice waves, as discussed in the preceding section, and imperfections of the lattice, and also a gradual loss of energy, due again mainly to the interaction with lattice waves. During this period, it is free to cover appreciable distances and, if an electric field is also present, will on the average carry a current in the field direction. If no other processes were possible, the electron and hole would ultimately end up in a state of very low velocity but would still contribute to the conductivity. In fact, since we have seen that the collision time varies as the inverse velocity, such slow carriers are particularly effective.

However, the life of these free carriers is, in fact, limited, since they are liable to other processes.

(a) *Recombination.* An electron may encounter a hole, and may then drop into the lower band, generally with the emission of radiation. It may alternatively lose, by collisions, enough energy while close to a hole to be confined to a closed orbit. The ultimate result will also be recombination.

(b) *Non-radiative recombination.* In some cases the recombination with the emission of radiation may be forbidden by the selection rules. This is important in particular if the lowest state in the upper band and the highest in the lower band do not belong to the same wave number. In this case the emission of radiation is still possible if simultaneously a phonon is produced or absorbed. This does not affect the energy balance appreciably, but may restore the balance of wave vectors. This, however, is a process of higher order and correspondingly less frequent.

Alternatively, the transition may take place without the emission of radiation, all surplus energy appearing in the form of phonons. Since the energy gap between the two bands is usually much larger than the energy of the hardest phonon, this usually means the simultaneous production of a large number of phonons. Such transitions are well known in the theory of molecular spectra, and the condition for them is that the potential energies of the two electronic states, plotted against the nuclear coordinates, should have an intersection. Fig. 16 represents such energy curves schematically, against some symbolic variable representing the motion of the nuclei in the lattice. If the curves intersect, we may regard the system as following the upper curve to the intersection and continuing on the lower curve with a corresponding increase of vibrational kinetic energy, which will be rapidly dissipated. However, the problem is essentially a many-body problem, so that such simple diagrams can only have a very qualitative meaning. A quantitative theory of this effect would have to be much more complicated.

(c) *Trapping.* Without returning to the lower band, the electron may lose its mobility by being trapped. One kind of trapped state is the one which we discussed in § 9.4, in which the lattice has distorted itself so as to conform to the forces due to an electron in a particular position, and thereby created an attractive centre for the electron. This process is referred to as 'self-trapping', and, while it seems theoretically likely that such states may exist, this form of trapping does not seem to have been definitely identified experimentally.

Alternatively, a trap may be some irregularity in the lattice which favours the presence of an electron. An example is a fault in an ionic crystal where a negative ion is missing, so that the neighbourhood contains too much positive charge (Mott and Gurney, 1940). It is likely that grain boundaries and other irregularities also can provide energy levels in which electrons can be trapped.

The same remarks apply to holes, which may also be trapped.

Trapped holes or electrons can be released by thermal energy fluctuations, which may give them enough energy to return to the mobile state, or by the absorption of light of much lower frequency than would be required to cross the gap between the full and the empty band.

This provides the mechanism for the most common type of phosphorescence. A 'phosphor' is a transparent crystal which contains a small number of impurity centres which absorb visible or ultra-violet light. The absorption process leads to a transition of an electron from an impurity centre into the empty band. Some of the electrons return to vacant impurity levels at once, with emission of radiation; others are trapped before they have found a vacant level. If the binding energy of the trap is comparable with kT at room temperature, they may after some time return to the empty band and then have another chance of radiating. The emission of light from such a phosphor continues therefore for some time after the irradiation has ceased. This process can be accelerated by heating, since this releases the electrons more frequently from the traps, and also releases those caught in deeper traps. The emission may also be accelerated by irradiation with infra-red light.

The frequency of the emitted light is usually less than that which excited the electrons in the first place. There are two reasons for this. Firstly, the electron may have been raised to a state above the bottom of the empty band but will in general lose energy by collision before it radiates. Secondly, when the electron is first removed from the impurity centre, the arrangement of the atoms surrounding this centre is very nearly the equilibrium arrangement corresponding to the electron being present. When it returns, the arrangement will be that most favourable for an ionized impurity atom, and when the electron has returned, the atoms will be in a configuration of higher potential energy, thus making less energy available for the transition. The remainder of the energy is, of course, then dissipated in the form of lattice waves.

Phosphorescence is of practical interest for producing luminous paints, which in the dark emit radiation as a result of the light they absorbed during daylight or when previously illuminated, and also for converting ultra-violet light into visible for greater efficiency and pleasanter appearance in fluorescent lighting.

Crystals may also be activated by fast electrons or other charged particles instead of light, and then provide a means of converting particles into light flashes, a possibility utilized in the scintillation counters of nuclear physics.

Because of these applications, as well as because of their intrinsic interest in the study of solids, the properties of luminescent substances have been studied very extensively (cf. Garlick, 1949).

XI

SUPERCONDUCTIVITY

11.1. Summary of properties (cf. Shoenberg, 1952)

In all other parts of this course we have been dealing with phenomena which were, at any rate qualitatively, well understood. This made it possible to present the theoretical treatment in a deductive manner, starting from general principles, although this did not reflect the actual historical development of the subject. In this chapter we are dealing with superconductivity, which cannot yet be claimed to be in a similar state. Within the last few years Fröhlich has proposed a picture of superconductivity which is very attractive, and very probably provides the right explanation, but the mathematical formulation is as yet incomplete, and it is therefore not certain that this model will lead to all the known features of superconductors without bringing in new effects that have at present been neglected.

I shall therefore start by outlining the main observed facts, and thus show the main results that must be derived from a complete theory of superconductivity, and I shall then review the theory of Fröhlich.

Superconducting metals have a transition temperature T_C, usually a few degrees, above which they behave like normal metals. As the temperature is lowered below T_C, the electrical resistance disappears suddenly. For pure metals the transition is extremely sharp, though inhomogeneities in structure or composition may spread it out.

Below the transition point the resistance appears to be zero, not merely very small. This was first demonstrated by inducing a current in a lead ring. The current continued for hours, in fact as long as it was practically possible to maintain the helium bath that kept the temperature low. This suggested that the state in which there was an electric current was a true equilibrium state which would last indefinitely, and this interpretation is borne out by the existence of the Meissner effect.

If an external magnetic field is applied to a superconducting ring, it will induce a current in the ring. The magnitude of this current is obvious from the fact that any change of the magnetic flux through the inside of the ring would result in an electromotive force in the metal. Since no electric field can exist in a substance of infinite conductivity, it follows that the magnetic flux encircled by the ring remains zero, and hence the flux due to the ring current must be equal and opposite to the applied field times the area of the ring.

Alternatively, if the ring is cooled in an external magnetic field, so that it passes into the superconducting state with a field present, the flux through the ring must remain constant from then on, and if the field is then removed, the flux is 'trapped' and there will be a current (equal and opposite to that of the previous experiment) to maintain this flux.

Meissner was the first to perform such experiments with solid objects, such as a superconducting sphere. Again, if the sphere is first cooled and a field is then applied, the lines of force do not penetrate into the material. The magnetic field inside the sphere therefore remains zero, and this means there must be surface currents flowing which compensate the external field. The sphere behaves like a diamagnetic body of permeability zero, or susceptibility $-1/4\pi$.

This much could have been foreseen from the infinite conductivity. However, if the sphere is cooled in the field, one finds that the final result is the same as before, i.e. surface currents start which reduce the field inside the sphere to zero.

This result, which is known as the Meissner effect, shows that the analogy between the superconducting sphere (or any other singly connected body) and a substance of zero permeability is complete, and that under changes of field and temperature the behaviour is reversible. This again shows that the state in which there are currents is a truly stationary state.

The irreversible behaviour of a ring is due to the fact that, once the ring is superconducting, no further changes of the flux through it are possible.

Superconductivity is destroyed when the field on the surface exceeds a certain critical strength. This critical field, which in general is of the order of a few hundred gauss, is a function of the temperature, and vanishes when the temperature reaches T_C. Its behaviour is approximately described by the relation

$$H_C = A(T_C^2 - T^2), \tag{11.1}$$

where A is a constant.

This also means that a superconducting wire is capable of carrying only a limited current, since stronger currents would produce fields which, on the surface of the wire, would exceed H_C.

The specific heat of a superconductor is discontinuous at the transition point, but, in the absence of a magnetic field, there is no latent heat.

§ 11.1   SUPERCONDUCTIVITY   213

If the transition takes place in a magnetic field, there is a finite latent heat.

These facts can be related to each other by applying thermodynamics. If the free energy per unit volume of the superconducting state without magnetic field is F_s, it will in a magnetic field be

$$F = F_s + \frac{1}{8\pi} H^2. \qquad (11.2)$$

This follows from the fact that the induced moment per unit volume, which is

$$M = -\frac{\partial F}{\partial H}, \qquad (11.3)$$

must equal $-H/4\pi$, to account for the Meissner effect.

At first sight one might believe that in the magnetic field the energy of a volume in which the field vanishes should be *lower* than that of a normal body because the term $H^2/8\pi$ in the electromagnetic energy density inside the body is missing. However, this reasoning neglects the fact that the flux which would have gone through the body must now take some other path, and therefore will increase the field energy elsewhere; if this is taken into account correctly, the result agrees with (11.2).

The transition to the normal state occurs when the free energy (11.2) equals that of the normal state F_n. Hence

$$F_s - F_n = -\frac{1}{8\pi} H_C^2, \qquad (11.4)$$

and by differentiation, using the usual thermodynamic relations,

$$S_s - S_n = \frac{1}{8\pi} \frac{d}{dT}(H_C^2), \qquad (11.5)$$

and

$$E_s - E_n = \frac{T^2}{8\pi} \frac{d}{dT}\left(\frac{H_C^2}{T}\right). \qquad (11.6)$$

The difference in the specific heats can be obtained by a further differentiation. These relations are well confirmed by experiment, thus justifying the assumption that we are dealing with equilibrium states.

Since H_C decreases with increasing temperature, the entropy and the energy content of the superconducting state are less than those of the normal state. At these low temperatures an important part of the entropy is that due to the conduction electrons, which is proportional to T. From the observed values of H_C, (11.5) contains a part proportional to T, which just cancels the electronic entropy of the normal state, leaving only a part proportional to a higher power of T, probably T^3. Thus the entropy of the electrons is reduced, and they are in a more ordered state.

The thermodynamic argument presupposes that as long as the metal is superconducting the magnetic field inside it necessarily vanishes.

In actual fact, the magnetic field must penetrate into a thin surface layer of the metal since the surface currents which shield the inside from the external field cannot be concentrated into a mathematical surface, but will spread over a layer of finite thickness. This penetration depth has been studied in many different experiments, and its magnitude is of the order of 10^{-5} cm.

The only other properties of the metal that show an anomalous behaviour at the transition point are the thermal conductivity, which, for a pure metal, is lower in the superconducting than in the normal state, and the thermoelectric effects, which disappear.

So far the main features to be explained by a theoretical model are:

(1) The existence of a transition to a state of lower entropy, and of permeability zero.
(2) The ability of the superconductor to carry circulating currents, if it is not singly connected, or to transmit currents without potential difference, if it is part of a circuit with normal components. It is often stated that (2) is a consequence of (1), but this has never been proved.
(3) Quantitative relationships, including the magnitude of the transition temperature and the critical field, as well as the penetration depth.

In a study of superconductors of different isotopic composition, it was discovered (Maxwell, 1950; Reynolds and others, 1950) that the transition temperature of a superconducting metal is different for different isotopes, and varies approximately as the inverse root of the nuclear mass. This shows directly that the mechanism of superconductivity must depend in some essential way on the motion of the atoms in the lattice, and this at once disproves all earlier theories which tried to explain the phenomenon in terms of the interaction of free electrons, or electrons in a rigid potential field, which would correspond to infinitely heavy atoms.

On the other hand, a new theory was put forward by Fröhlich just before the discovery of the isotope effect, in which the main mechanism is the interaction between electrons and lattice waves. It is the only type of theory which can account for the isotope effect in a natural manner.

11.2. Outline of the Fröhlich–Bardeen theory

According to Fröhlich (1950), superconductivity is due to the interaction of the electrons with the lattice waves which, indirectly, results in an interaction between the electrons.

Consider first a single electron with wave vector **k** in a lattice in which there are no phonons. The interaction with the lattice vibrations which was considered in Chapter VI then gives rise to the possibility of creating a phonon, with a corresponding change in the electron state to **k**′. If in this transition energy is conserved, it becomes a real transition of the kind that we found important for the resistance of an ideal metal. Even where such real transitions are not possible, however, transitions in which energy is not conserved must still be allowed for as virtual transitions, i.e. the electron may spend a part of its time in the state **k**′, and is then accompanied by a photon.

Such virtual transitions always result in a perturbation of the energy, the change in the energy of the electron state being

$$\Delta E(\mathbf{k}) = -\frac{\hbar}{2\pi} \sum_{\mathbf{k}',s} \frac{|(\mathbf{k}'|A|\mathbf{k},s)|^2}{E(\mathbf{k}')+\hbar\omega(\mathbf{f},s)-E(\mathbf{k})}, \quad (11.7)$$

where the matrix element in the numerator is given by (6.48), and **f** is the wave vector of a phonon satisfying the conservation law (6.46). I have considered only transitions within the same band and omitted the band suffix l for simplicity.

Equation (11.7) is the standard result for second-order perturbation theory. The terms in the sum for which the denominator is positive correspond to virtual transitions to states in which the sum of phonon and electron energy is higher than $E(\mathbf{k})$, and they tend to make the perturbation energy negative, whereas terms with negative denominators raise the perturbed energy.

A vanishing denominator corresponds to a real transition. If such terms occur, the expression (11.7) is ambiguous, but one knows from general theory that one should interpret the sum by imagining that the denominator has an infinitesimal negative imaginary part. On replacing the summation by an integration, the result then may be expressed as the principal value of the integral, which is real and finite, plus one-half the residue at the pole, which gives an imaginary contribution. An imaginary part of the energy of the state **k** represents an exponential decay with time, and therefore expresses the fact that, because of real collisions, the electron only spends a finite time in the state **k**.

In discussing the properties of the lowest state of the whole system, such real collisions must, however, be unimportant, since in the lowest state no energy is available for the emission of a phonon.

As it stands, the energy perturbation (11.7) is of no great interest since it amounts merely to a very small correction to the electronic energy function, which, in any case, is not known to high accuracy.

However, as Fröhlich noticed, the same effect causes an interaction between the electrons. If several electrons are present, we must omit from the sum in (11.7) any states $\mathbf{k}'$ which are already occupied, since by Pauli's principle an electron cannot make a transition, either real or virtual, to an occupied state. Hence the energy correction for all electrons becomes, in this approximation,

$$\frac{\hbar}{2\pi} \sum_{\mathbf{k},\mathbf{k}',s} \frac{|(\mathbf{k}'|A|\mathbf{k},s)|^2 n(\mathbf{k})\{1-n(\mathbf{k}')\}}{E(\mathbf{k})-E(\mathbf{k}')-\hbar\omega(\mathbf{f},s)}, \qquad (11.8)$$

where $n(\mathbf{k})$ is the occupation number.

We now divide this expression into two parts. The part due to the 1 in the numerator is just the sum of (11.7) over all occupied states, and is therefore equivalent to a small adjustment in the energy function. This leaves

$$-\frac{\hbar}{2\pi} \sum_{\mathbf{k},\mathbf{k}',s} \frac{|(\mathbf{k}'|A|\mathbf{k},s)|^2 n(\mathbf{k})n(\mathbf{k}')}{E(\mathbf{k})-E(\mathbf{k}')-\hbar\omega(\mathbf{f},s)}. \qquad (11.9)$$

This expression may be written in a more symmetrical form by combining each term with the one in which $\mathbf{k}$ and $\mathbf{k}'$ are interchanged. These two terms have the same matrix element, and belong to opposite phonon wave vectors, which have the same frequency. This gives

$$-\frac{\hbar}{2\pi} \sum_{\mathbf{k},\mathbf{k}',s} \frac{\hbar\omega(\mathbf{f},s)|(\mathbf{k}'|A|\mathbf{k},s)|^2 n(\mathbf{k})n(\mathbf{k}')}{\{E(\mathbf{k}')-E(\mathbf{k})\}^2-\{\hbar\omega(\mathbf{f},s)\}^2}. \qquad (11.10)$$

The value of this expression is now sensitive to the distribution of the electrons. Two electrons of the same energy, but moving in different directions, contribute to (11.10) a positive term. However, if their energy is slightly different, by just the amount which corresponds to a real transition, the denominator changes sign. For all greater energy differences the denominator remains positive, thus giving a reduction of the energy.

This interaction therefore favours electron arrangements in which fewer electrons occupy immediately adjacent points in $\mathbf{k}$ space, and more of them are separated by distances sufficient to make the denominator of (11.10) negative.

Fröhlich has evaluated the integral arising from (11.10) for the case in which the electrons may be regarded as free, and the lattice waves as those of an isotropic medium. He calculates the interaction energy for a state in which the electrons occupy a sphere about the origin in **k** space, and in addition a concentric shell, with the inner radius of the shell larger than that of the sphere by a gap a, equal to the thickness of the shell. For $a = 0$, this reduces to the solid sphere, and hence gives the Fermi distribution at $T = 0$. Fröhlich finds that the interaction energy is lower for a finite gap width than for $a = 0$. The most favourable value of a corresponds to energy differences of the same order as found in electron-phonon collisions, and the ratio of the width a of the gap to the radius of the sphere is therefore of the order of the ratio of the velocity of sound to the electron velocity, i.e. of the order of 10^{-3}.

Hence the magnitude of the energy reduction can be estimated in the following way: the denominator in (11.10) is of the order of the square of a phonon energy; in the numerator the first factor is a phonon energy, and the square of the matrix element (including the factor $\hbar$) is, by the estimate (6.62), of the order of D, where D is an electronic energy. The number of terms making large contributions to (11.10) is given by the number of pairs of states which lie within the shell or within a similar surface layer of the main sphere. This makes the number proportional to $N^2(\hbar\omega)^2/D^2$, and the reduction in interaction energy of the order

$$\frac{N(\hbar\omega)^2}{D}. \tag{11.11}$$

On the other hand, the unperturbed electron energy is lowest for the usual Fermi distribution, i.e. for the solid sphere, and by creating the shell it is increased by an amount which is again of the order of (11.11), since we raised $N\hbar\omega/D$ electrons by energy amounts of the order $\hbar\omega$.

According to whether the change in the interaction energy or in the unperturbed energy is the greater, the sphere with shell or the solid sphere represents the state of lowest energy. It is not known whether it would be possible to construct a state of still lower energy, but the order of magnitude of the reduction would not exceed (11.11).

This argument suggests that certain metals have at $T = 0$ a state in which the electron distribution is rather different from the usual Fermi model. As the temperature is raised, the distribution remains at first similar, but the number of low excited states of the metal will be less than usual, since the energy is now very sensitive to the details of the electron distribution, so that a slight displacement of an electron in k space is of importance. It is therefore reasonable to expect the entropy of this state to be lower than that of the normal state;

as the temperature rises there will come a point where the normal state is stable. This question has not yet been investigated in detail.

The quantity (11.11) is certainly of the same order of magnitude as the observed energy difference (11.6) at $T = 0$. Since the electronic specific heat of the superconducting state is at low temperatures much less than that of the normal state, and since there is no latent heat at the transition point, the electronic energy content of the normal state at T_C should approximately equal the energy difference at $T = 0$. Using (11.11) for the latter, this gives also the right order of magnitude for T_C.

If we compare different isotopes, the quantity (11.11) is proportional to the inverse of the atomic mass, since the vibration frequencies vary inversely as the root of the mass, and since the thermal energy of the electrons in the normal state varies as T^2, this makes T_C proportional to the inverse square root of the mass. This result is compatible with the observed isotope effect, though the power law $M^{-\frac{1}{2}}$ is not yet established with accuracy.

A theory based on essentially the same mechanism, but using a different method of approximation, has been given by Bardeen (1950) (see also Bardeen, 1951). In this the use of the perturbation formula (11.7), which for small energy denominators is very questionable, is avoided by using instead a variational method. The results are very similar to Fröhlich's. This method, however, proceeds by constructing a separate wave function for each electron in interaction with the lattice vibrations, using wave functions which are not orthogonal to each other; it also cannot be claimed to be rigorous.

11.3. Effect of a magnetic field

A suggestion for describing the Meissner effect is due to London (1948). The interaction of an electron with a magnetic field of axial symmetry may be written (cf. (7.10))

$$\frac{e}{mc}Ap_\phi + \frac{e^2 A^2}{2mc^2}, \qquad (11.12)$$

where r is the distance from the axis, the vector potential has a component only in the azimuthal direction, and p_ϕ is the angular momentum about the axis. If the potential in which the electron moves has axial symmetry, then p_ϕ has discrete values. If we fix our attention on a state with $p_\phi = 0$, or on a number of electrons for which the average value of p_ϕ vanishes, then their energy in the field is increased by the second term. If their density n in space is approximately constant, the total

energy density would then depend on the field by a term
$$\frac{ne^2 A^2}{2mc^2},\qquad(11.13)$$
and the current density would be
$$j = -\frac{\partial E}{\partial A} = -\frac{ne^2 A}{mc^2}.\qquad(11.14)$$
If this result is inserted in Maxwell's equations, one finds that the vector potential decreases exponentially from the surface of the body inwards, the mean penetration depth being of the order
$$\left(\frac{mc^2}{4\pi ne^2}\right)^{\frac{1}{2}}.$$

This consideration is used in an attempt by Fröhlich (1951) to show that his model leads to a result similar to (11.14). He argues that in a normal metal the term (11.13) is compensated by the fact that electrons are transferred from states of positive p_ϕ to states of negative p_ϕ even though, to satisfy the Pauli principle, this requires an increase in their kinetic energy. The net result is just to cancel the energy change, except for the small diamagnetism of a normal metal. If the electron energy is more sensitive to the electron distribution, it would require more energy to redistribute the electrons, and therefore the term (11.13) would no longer cancel exactly, and a fraction of the current (11.14) would remain, thus giving a Meissner effect with a somewhat greater penetration depth.

This argument overlooks the fact that the interaction of the electrons with the lattice waves no longer leaves the angular momentum of the electrons constant; the first term in (11.12) therefore is no longer just a constant energy shift for each state, but causes a perturbation which gives also second-order contributions proportional to A^2. It would have to be shown that these do not cancel the remainder of (11.13).

An alternative explanation is proposed by Bardeen (1950) who regards the Meissner effect as the case of a strong electronic diamagnetism. Since, according to Chapter VII, the diamagnetic susceptibility is inversely proportional to the square of the effective mass of the electrons, a decrease of this mass by a factor of the order 10^3 would make the susceptibility of the order of $-1/4\pi$ as required for the Meissner effect. Since on the present model of superconductivity it takes much more energy than usual to change the state of an electron, this is equivalent to a very small effective mass. However, the use of the simple formula for the susceptibility in such circumstances is not justified, since the induced moment is proportional to the actual field at each point. In the language of the phenomenological equations, the induced moment should in this case be expressed as χB, rather than χH. In the case of ordinary diamagnetism, this difference is unimportant, but it follows that a negative infinite χ would be required to make the permeability zero.

This also follows from the fact that otherwise an even smaller effective mass might make the permeability negative, which would not make sense.

11.4. Objections and difficulties

In spite of the many encouraging features of the model, there remain a number of serious difficulties.

(1) The use of perturbation theory is very doubtful when the perturbation is large enough to reverse the order of different states of the system. This is particularly serious when one is dealing with a continuous spectrum for each electron. The expression (11.9) may be interpreted as a collision in which the virtual phonon emitted by one electron is absorbed by another in such a manner that the electrons exchange their wave vectors. Because of the identity of the electrons, this still leaves the system in the original state. Evidently, however, a collision in which the electrons exchange a phonon and both change their states in a more general way may still conserve energy, and therefore is of equal importance. To take such terms into account one would, however, have to start from a wave function which allows for correlations between the electrons. Even the variation method of Bardeen, which avoids perturbation theory to some extent, does not seem capable of handling this problem.

A more recent paper by Fröhlich (1952) substitutes a canonical transformation for perturbation theory. Since, however, a great number of approximations are made in carrying this out, it is not yet clear whether the result is, in fact, more accurate.

(2) According to Wentzel (1951), the same interaction which leads to (11.7) also leads to a dependence of the perturbed energy on the number of phonons present. If the coupling is strong enough to satisfy Fröhlich's criterion for superconductivity, the perturbation decreases sufficiently strongly with phonon number to outweigh the unperturbed phonon energy. In other words, the energy of a phonon would now be negative, implying an instability of the lattice.

For a one-dimensional model this result was obtained by Wentzel without using perturbation theory. There is a close connexion between this result and the remark in § 5.3 that any deviation from perfect translational symmetry of a linear chain of unsaturated atoms must lead to a decrease of energy. The difference between these points of view is that in § 5.3 we were discussing the effect of a static potential, whereas

Wentzel's argument allows for the motion of the atoms. In the simpler case of atoms at rest, it is clear that, as in Jones's theory of bismuth, the effect will lead merely to a new equilibrium configuration of a less regular structure, and it is likely that the same would happen in Wentzel's model if one were to treat the forces between the atoms more exactly than to the harmonic approximation. However, one would then not be justified in using a treatment in which the atomic displacements from the regular structure were regarded as small.

We saw in § 5.3 that this effect tended to alter the band structure in such a way as to have the actual electron gas just fill one band completely, and at first sight this suggests that the resulting state of lower entropy should have a poor conductivity rather than superconductivity. It is therefore particularly important to examine carefully whether the 'superconducting' state of the Fröhlich model really possesses the electric and magnetic properties of a superconductor.

(3) The model results in an indirect interaction between electrons, with phonons as intermediaries. This is a very weak interaction, and therefore does fit in with the very low transition temperatures and small energy differences observed in practice. However, it seems inconsistent to include this small effect and neglect other causes of interaction, in particular the electrostatic interaction between the electrons, which also produces correlations between them, unless one can see that the indirect interaction has qualitative consequences which are insensitive to the presence of electrostatic forces.

In fact, Heisenberg (1947) had proposed an alternative theory based entirely on the electrostatic interaction, which also seemed capable of producing a change of state at very low temperatures. The method used is perturbation theory of a type similar to that of the preceding section, and therefore open to the same objections. From the isotope effect, it is now clear that the electrostatic effect, which is independent of the motion of the nuclei, cannot by itself be the cause of superconductivity, but its influence on the indirect interaction remains to be discussed.

Wentzel suggests, in fact, that the electrostatic interaction is important to correct the lattice instability to which he has drawn attention, but it is doubtful whether it is necessary for this purpose, and whether it could do this.

These brief remarks on the most fascinating problem of the theory of solids will have to suffice. With the isotope effect as clue, and the Fröhlich–Bardeen model as a starting-point, further progress may be expected in this field.

BIBLIOGRAPHY

REFERENCES are given only to papers in which the reader can find more detail or more complete arguments, and no claim is made to give the original sources of each idea.

For a more complete introduction to large parts of the subject the following books and articles may be used:

KITTEL (1953)
SEITZ (1940)
MOTT and JONES (1936)
BETHE and SOMMERFELD (1933)
WILSON (1936)

In addition to these, for further information on the subject matter of each chapter, the reader is referred to:

Chapter I BORN (1923)
PEIERLS (1934)
I to III BORN and HUANG (1954)
III JAMES (1948)
IV BRILLOUIN (1946)
VII STONER (1947)
VIII STONER (1947)
IX MOTT and GURNEY (1940)
X GARLICK (1949)
SHOCKLEY (1950)

REFERENCES

ADAMS, 1953, *Phys. Rev.* **89**, 63. — page 155
BARDEEN, 1950, ibid. **79**, 167, and **80**, 567. — 218, 219
—— 1951, *Rev. Mod. Phys.* **23**, 261. — 218
BERMAN, 1951, *Proc. Roy. Soc.* A, **208**, 90. — 43
—— 1953, *Phil. Mag. Suppl.* **2**, 103. — 43
BETHE and SOMMERFELD, 1933, *Handbuch der Physik*, vol. **24/2**, p. 333, Springer. — 169, 222
BLACKMAN, 1938, *Proc. Roy. Soc.* A, **166**, 1. — 154
—— 1942 *Phys. Soc. Reports on Progress in Physics*, **8**, 11. — 31
BLOCH, 1928, *Z. Physik*, **52**, 555. — 75
—— 1929, ibid. **57**, 545. — 174, 177
BORN, 1923, *Atomtheorie des festen Zustandes*, Teubner. — 222
—— 1943, *Phys. Soc. Reports on Progress in Physics*, **9**, 294. — 67
—— and HUANG. 1954, *Dynamical Theory of Crystal Lattices*, Oxford. — 222
BRILLOUIN, 1922, *Ann. de Physique*, **17**, 88. — 70
—— 1931, *Quantenstatistik*, Springer. — 222
—— 1946, *Wave Propagation in Periodic Structures*, McGraw-Hill. — 222
CARR, W. J., jr. 1953, *Phys. Rev.* **92**, 28. — 183
CASIMIR, 1938, *Physica*, **5**, 495. — 53
DINGLE, 1952, *Proc. Roy. Soc.* A, **211**, 517. — 155
DRUDE, 1904, *Ann. Physik*, **14**, 936. — 184
FRENKEL, 1936, *Phys. Zeits. d. Sowjetunion*, **9**, 158. — 194
FRÖHLICH, 1950, *Phys. Rev.* **79**, 845. — 215
—— 1951, *Proc. Phys. Soc.* A, **64**, 129. — 219
—— 1952, ibid. **215**, 291. — 220
GARLICK, 1949, *Luminescent Materials*, Oxford. — 210, 222
DE HAAS and VAN ALPHEN, 1930, *Proc. Amsterdam Acad.* **33**, 1106. — 154
—— and BIERMASZ, 1935, *Physica*, **5**, 495. — 53
HARPER, 1954, Ph.D. thesis, Birmingham. — 151, 154
HEISENBERG, 1947, *Z. Naturforschung*, **2**, 185. — 221
HELLMAN, 1933, *Z. Physik*, **85**, 180. — 102
JAMES, 1948, *Optical Principles of Diffraction of X-rays*, London. — 222
KELLERMAN, 1940, *Phil. Trans. Roy. Soc.* A, **238**, 513. — 31, 57
KITTEL, 1953, *Introduction to Solid-State Physics*, New York. — 222
KLEMENS, 1951, *Proc. Roy. Soc.* A, **208**, 108. — 52
KRAMERS, 1934, *Physica*, **1**, 184. — 182
LONDON, 1948, *Phys. Rev.* **74**, 562. — 218
MARSHALL, 1954, *Proc. Phys. Soc.* A, **67**, 85. — 172
MAXWELL, 1950, *Phys. Rev.* **78**, 447, and **79**, 173. — 214
MITCHELL, 1934, *Proc. Roy. Soc.* A, **146**, 442. — 192
—— 1936, ibid. **153**, 513. — 192
MOORHOUSE, 1953, ibid. **64**, 1097. — 177
MOTT, 1936, *Proc. Camb. Phil. Soc.* **32**, 281. — 125
—— 1938, ibid. **34**, 568. — 202
—— 1939, *Proc. Roy. Soc.* A, **171**, 27. — 205
—— and GURNEY, 1940, *Electronic Processes in Ionic Crystals*, Oxford. — 209, 222
—— and JONES, 1936, *Properties of Metals and Alloys*, Oxford. — 113, 139, 144, 164, 272

REFERENCES

NÉEL, 1932, *Ann. de Physique*, **17**, 64. *page* 181
—— 1936, ibid. **5**, 256. 181
NORDHEIM, 1931, *Ann. Physik*, **9**, 641. 125
PAULI, 1926, *Z. Physik*, **41**, 81. 144
PEIERLS, 1929, *Ann. Physik*, **3**, 1055. 222
—— 1933, *Z. Physik*, **80**, 763, and **81**, 186. 151, 152
—— 1934, *Ann. Inst. H. Poincaré*, **5**, 177. 222
—— 1934 a, *Helv. Phys. Acta*, **7**, suppl. 2, 24. 140
PLACZEK, 1934, *Handbuch d. Radiologie*, 6/2, 209. 70
—— 1952, *Phys. Rev.* **86**, 377. 70
POMERANCHUK, 1943, *J. of Phys. U.S.S.R.* **7**, 197. 51
REYNOLDS and others, 1950, *Phys. Rev.* **78**, 487. 214
SCHIFF, 1949, *Quantum Mechanics*, McGraw-Hill. 40, 55
SCHUBNIKOW and DE HAAS, 1930, *Proc. Amsterdam Acad.* **33**, 418. 160
SEITZ, 1940, *Modern Theory of Solids*, McGraw-Hill. 222
SHOCKLEY, 1950, *Electrons and Holes in Semi-Conductors*, van Nostrand. 222
SHOENBERG, 1939, *Proc. Roy. Soc.* A, **170**, 341. 154
—— 1952, *Superconductivity*, Cambridge. 212
—— 1952 a, *Phil. Trans. Roy. Soc.* A, **245**, 1. 154
SKINNER, 1938, *Phys. Soc. Reports on Progress in Physics*, **5**, 257. 189
SLATER, 1951, *Quantum Theory of Matter*, McGraw-Hill. 5
SLONIMSKI, 1937, *J. Exp. and Theor. Phys. U.S.S.R.* **7**, 1457. 53
STONER, 1947, *Magnetism*, Methuen. 222
—— 1948, *Phys. Soc. Reports on Progress in Physics*, **11**, 43. 177
TITCHMARSH, 1937, *Introduction to the Theory of Fourier Integrals*, Clarendon Press.
148
TITEICA, 1935, *Ann. Physik*, **22**, 129. 160
VAN VLECK, 1951, *J. Physique et Radium*, **12**, 262. 183
WENTZEL, 1951, *Phys. Rev.* **83**, 168. 220
VAN WIERINGEN, 1954, *Proc. Phys. Soc.* A, **67**, 206. 141
WILSON, 1936, *Theory of Metals*, Cambridge. 152, 187
—— 1939, *Semi-Conductors and Metals*, Cambridge. 158
ZENER, 1933, *Nature*, **132**, 968. 186

LIST OF SYMBOLS

Symbols are not listed here if they occur only in immediate context with a statement defining them. Conventional mathematical symbols are not listed. References to defining equations are given only when a verbal statement of the meaning is inconvenient, or inadequate.

Symbol	Meaning	Page	Equation	Used only in
a	lattice spacing	1		
$\mathbf{a_n}$	lattice vector	4		
$A_{n,n'}$	coefficient of harmonic term (linear chain)	11		
$A_{j,n;j',n'}$	coefficient of harmonic term (general)	14		
$\mathbf{A}$	vector potential			
$(...|A|...)$	scattering matrix element	126		
$b(...)$	coefficient of cubic term in terms of normal coordinates	36		
$b(...)$		38	(2.49)	
$B_{n,n',n''}$	coefficient of cubic term (linear chain)	11		
$B_{j,n;j',n';j'',n''}$	coefficient of cubic term (general)	14		
c	velocity of sound			
c	velocity of light			Chapters III, VII, X
$c(...)$	coefficient of quartic term in terms of normal coordinates	39		
$c(...)$		39	(2.54)	
C	reduced interaction constant	130		
$\mathbf{d}_j$	position of atom in cell	4		
D	order of magnitude of band width	130		Chapter VI
D	diffusion coefficient			Chapter X
e	electron charge			
e_j	charge of ion			
E	energy			
$E_l(\mathbf{k})$	energy of electron state	77		
$\mathbf{E}$	electric vector			§ 9.1
$\mathbf{f}$	wave vector of lattice wave	15		
$f(E)$	Fermi function	91		
F	free energy			
F	form factor			§ 3.2
$\mathbf{F}$	electric field intensity			
$g(...)$	distortion of phonon distribution	49		Chapter II
$g(...)$	distortion of electron distribution	128		Chapter VI
$G(...)$	distortion of phonon distribution	128		Chapter VI
G	Fourier transform of force constants	15		
$\mathbf{g}$	wave vector of spin wave	168		
$\hbar$	$= h/2\pi$			
h	Planck's constant			
$\mathbf{H}$	magnetic field intensity			
I	electron interaction	132		
j	label of site in cell	4		
$\mathbf{J}$	total wave vector	42		Chapter II
$\mathbf{J}$	current density	118		Chapters VI, VII
k	Boltzmann's constant			
$\mathbf{k}$	light or neutron wave vector	55		Chapter III, § 8.5
$\mathbf{k}$	electron wave vector	77		
$\mathbf{K}$	reciprocal lattice vector	16		
L	linear dimension of crystal	15		
l	band label	77		

LIST OF SYMBOLS

Symbol	Meaning	Page	Equation	Used only in
m	electron mass			
m^*	effective mass	90		
M	atomic mass	11		
$M^{(N)}$	mass of crystal	19		
M^C	mass of unit cell	19		
n	cell label	1		
$n_l(\mathbf{k})$	number of electrons per state	117		
N	total number of cells			
$N(...)$	phonon number	25		
$\mathbf{p}$	electron momentum	83		§4.3
q	normal coordinate	17		
$\mathbf{q}$	wave vector difference	59		
Q	normal coordinate of progressive wave	17		
$\mathbf{r}$	general radius vector			
r	number of atoms per cell	4		
$\mathbf{R}$	position of atom			
R	Hall constant			§§7.4, 10.3
s	type of lattice wave	17		
S	energy transport	120		
T	kinetic energy			§1.7
T	temperature			except in §1.7
$\mathbf{u}$	atomic displacement	11		
$u_{\mathbf{k},l}$	wave function without phase factor	77	(4.9)	
U	potential energy	7		
$U(\mathbf{r})$	atomic potential	79		
V	volume of crystal	16		
$V(\mathbf{r})$	potential energy of electron	76		
$\mathbf{v}$	electron velocity	87		
$v_j(\mathbf{f}, s)$	relative amplitude in normal mode	17		
W	work function	97		
x, y, z	coordinates			
Z	partition function	35		Chapter II
Z	number of electron states	93		except Chapter II
α	interaction integral	166		
β	$= 1/kT$			
β	overlap integral	166		Chapter VIII
γ	exchange integral	166		Chapter VIII
$\delta(x)$	Dirac delta function			
δ_{ik}	Kronecker symbol			
Δ	generalized Kronecker symbol	38	(2.50)	
ϵ	polarization vector			
η	Fermi energy	92		
Θ	Debye temperature	30		Chapters II, VI
Θ	Curie temperature	162		Chapter VIII
κ	thermal conductivity			
λ	mean free path			
μ	Bohr magneton			
μ	component of electron spin	165		§8.2
ν	general label for atomic coordinate	19		
ρ	density			
ρ	surface density of electron states	116		§6.1
σ	label for progressive lattice waves	17		
σ	electric conductivity			
τ	collision time	116		
ϕ	scalar potential			§10.4
ϕ	general label for normal coordinate	19		Chapter I
$\phi(\mathbf{r})$	atomic wave function			§4.2, Chapter VIII
ψ	electron wave function	76		
ω	frequency			
$\omega(\mathbf{f}, s)$	frequency of lattice wave	17		
Ω	Larmor frequency	145		Chapter VII

INDEX

(*A bold-face page number indicates the main reference or the explanation of the term.*)

absorption: of infra-red radiation, 54; of light, 188; of X-rays, 188.
— edge, 192.
acceleration, 87, 156.
acceptor, 197.
acoustical branch, 30, 56.
adiabatic approximation, 4, 111.
alkali halides, 21; *see* NaCl type lattice.
alkalis, 6, 138, 159.
anharmonic terms, 31.
anti-ferromagnetism, 181.
anti-Stokes line, 69.

background scattering, 64.
band, 81, 86, 89, 197.
basic cell (of reciprocal lattice), 17, 36, 77.
bismuth, 113, 152.
Bloch theorem, 75, 126, 130, 131.
blocking layer, 205.
body-centred cubic lattice, 2, 16, 82, 104, 171.
Bohr magneton, 143.
Boltzmann equation, 45, 117, 127, 156, 160.
boundary region, 121, 132.
Bragg's law, 61, 84, 87, 112.
Brillouin zone, 77, 87.
bulk modulus, 24, 32.

centre of symmetry, 7.
characteristic temperature, 30, 133, 135.
coherent scattering, 59.
collective electron model, 173, 178.
collision time, 115, 116, 120, 133, 139, 156, 184, 201.
collisions: between electrons, 131, 138; between electrons and phonons, 126, 136, 217; between phonons, 41, 46, 53, 133, 136.
commutation law, 24, 151.
compressibility, 24, 101.
conservation laws, 43, 128.
contact potential, 97.
cores, 96, 102.
correlation, 103, 106, 176.
critical field, 212.
cubic terms, 11, 32.
Curie point, 162, 173.
cyclic condition, 13, 15, 76, 146.

Debye model, 29, 33, 136.
Debye-Scherrer method, 61.
Debye temperature, *see* characteristic temperature.
degenerate electron levels, 76, 81.
— Fermi gas, 92, 144, 200.
density matrix, 47.
detailed balancing, 116.
diamagnetic susceptibility, 149, 152, 219.
diamagnetism, 144.
diamond, 6, 10.
dielectric constant, 185.
diffraction of X-rays, 58, 112.
diffusion coefficient, 204.
distorted structure, 108, 112.
domain, 161, 179.
donor, 197.
double layer, 96, 104.
drift of phonons, 137.
Dulong-Petit, 30.
duration of collision, 124.
dynamical theory, 63.

effective mass, 91, 106, 152, 199, 219.
Einstein relation, 204.
elastic constants, 22, 101, 108.
electric conductivity, 118, 134, 136.
— — of semi-conductors, 199.
— current, 118.
— field, 88, 120.
electron diffraction, 86.
electrostatic forces between ions, 9, 57.
emission of X-rays, 189.
energy of vibration, 20, 25.
— surface, 116.
— transfer in collisions, 135.
— transport, 42, 51, 120.
entropy, 213, 217.
exchange, 163, 182.
— integral, 167, 174, 181.
exciton, 194.

face-centred cubic lattice, 2, 16, 104, 171.
Fermi-Dirac statistics, 91.
Fermi energy, 95, 97, 199.
— function, 92.
— surface, 113, 138.
ferromagnetic susceptibility, 180.
ferromagnetism, 161.
form factor, 59, 62.

INDEX

Fourier coefficients, 63.
four-phonon processes, **43**, 51.
fourth-order terms, **36**, 39, 43, 51.
free electrons, **83**, 118, 186, 192.
— energy, 31, 145, 213.
frequency of lattice vibrations, 13, 17, 28.
Fröhlich theory of superconductivity, 114, **215**.

graphite, 10.
Grüneisen's law, **33**.

de Haas–van Alphen effect, **154**.
Hall effect, **155**, 202.
harmonic oscillator, **35**.
Hartree method, 102.
Heitler–London model, **164**, 193.
hexagonal close-packed lattice, 3, 32.
hole, **91**, 159, 193.
homopolar bond, 6, **10**.
hysteresis, 180.

imperfection and impurities, 40, 52, 115, 121, 134.
impurity semi-conductor, **198**.
infra-red absorption, 54.
insulator, **89**, 197.
interaction between electrons, 89, 103, 115, **132**, 164, 173, 198, 221.
ionic lattice, 6, 9, 31, 54, 194.
isotope effect, **214**, 218.

Jones theory of Bi, **113**, 155, 221.
Joule heat, 117.

Larmor frequency, 145, **157**.
lattice, **1**.
— vector, **1**.
Laue method, **61**.
light waves, 26, 184.
linear chain, 11, 67, 79, 83, 108, 165, 183.
longitudinal waves, 21, 43, 44, 51.
Lorentz-Lorenz factor, 187.

magnetic field, 88, 143.
magnetization, 179.
magneto-resistance, **155**.
mean free path, 40, 52, 116, 202.
Meissner effect, **211**, 218.
metal, 6, 75, 89, 197.
metallic bond, 11, 101.
mobility, **158**.
molecular field, **161**.
— solid, 6, 31.
mosaic structure, 8.

NaCl type lattice, 3, 32, 192.
neutrons, 70, 177.
non-diagonal terms, **151**, 154.

non-metal, 6, 75.
normal coordinates, **14**, 15, 36, 126.
— vibrations, **12**, 15, 17.

optical branch, **30**, 56.
orthogonality, **18**.
overlap, 10, 166, 182, 183.

paramagnetism, **143**.
Pauli principle, 10, 89, 102, 103, 116, 216.
penetration depth, **214**.
periodic potential, 75.
perturbation theory, 40, 80, 83, 124, 141, 215, 220.
phase relations, 47, 123.
phonon, **26**, 41, 54.
phosphor, **209**.
photoelectric effect, **190**.
photon, 55, 70, 188, 191.
positron theory, 91.
potential barrier, 83, 99.

quantum number, 25.
quartic terms, *see* fourth-order terms.

Raman effect, 55, 70.
reciprocal lattice, **16**, 36, 41, 60, 77, 126.
recombination, 208.
rectification, **205**.
rotating crystal method, 61.

scattering: of light, 55, 68; of neutrons, 70, 177; of phonons, 52.
screening, 131.
semi-conductor 76, **197**.
shear, 7.
simple cubic lattice, **1**, 16, 171.
— hexagonal lattice, **2**.
size effect, 53.
sound velocity, 29, 40.
— waves, 21, 26, 53.
specific heat: of lattice, **27**, 34; of electrons, 93; of superconductor, **212**.
spin, 89, 143, 161.
— complex, **169**.
— waves, **164**, 178.
stability of equilibrium, 7, 220.
standing waves, 85.
Stokes line, **69**.
Stosszahl-Ansatz, **123**.
strain, 24, 34.
strongly bound electrons, 79, 164.
superconductivity, **211**.
surface, **95**, 187, 190, 193.
— waves, **100**.

temperature gradient, 46, 120.
thermal conductivity: of electrons, **120**, 138; of lattice, 40, 45, 128.
— expansion, 31, 39.

INDEX

thermionic emission, **98**.
transition probability, 40, 55, 115, 122, 126, 189.
— temperature, **211**.
translational symmetry, 75.
translation group, **1**.
transverse waves, 21, 43, 44.
trapping, 196, 209.
two-dimensional crystal, 67, 183.

Umklapp process, **41**, 50, 53, 137, 139.
uncertainty principle, 45.

variation principle, 80.

velocity of electron, **87**.
volume effect, 192.

van der Waals forces, 9.
wave number, **13**.
— vector, **15**, 43, 126, 188, 193.
— packet, 45, 87, 120, 156.
Weiss model, **161**, 168, 181.
Wiedemann-Franz law, **121**, 134.
Wigner–Seitz method, **104**, 189.
work function, **97**.

zero-point energy, 26, **27**, 135.